Clyde Tombaugh
DISCOVERER OF PLANET PLUTO

Clyde Tombaugh

DISCOVERER OF PLANET PLUTO

DAVID H. LEVY

Sky Publishing Corp.
Cambridge, Massachusetts

For Mom and Dad — I'll always remember
when they first told me how Pluto was found.

© 2006 David H. Levy
Published by Sky Publishing Corporation
49 Bay State Road
Cambridge, MA 02138-1200, USA
SkyandTelescope.com

Library of Congress Cataloging-in-Publication Data

Levy, David H., 1948-
    Clyde Tombaugh : discoverer of planet Pluto / David H. Levy.
        p. cm.
    Originally published: Tucson : University of Arizona Press, c1991.
    Includes bibliographical references and index.
     ISBN 1-931559-33-3 (alk. paper)
    1. Tombaugh, Clyde William, 1906-1997. 2. Pluto (Planet)
    3. Astronomers--United States--Biography. I. Title.

QB36.T6L48 2006
520.92aB--dc22

2006004021

CONTENTS

NEW PREFACE

I MISS Clyde.

From the time I first heard of him, during the summer of 1960, I admired the man and what he stood for. That's when Dad told me about the outer planets of our solar system: how Herschel's discovery of Uranus led to Neptune, and how Neptune led to the search — Clyde's search — that revealed the existence of Pluto.

I miss our decades of friendship. I first enountered Clyde at the Symposium on the Exploration of Mars in Denver in 1963. I was 15. After a panel discussion the meeting was thrown open for questions. Someone approached the microphone. "Clyde Tombaugh," he identified himself. I turned to face him in shock — I was in the same room as the discoverer of Pluto!

I miss the man of multiple achievements in science, engineering, and even literature. For example, here's a poem called "Our Team" that he wrote about 1921, while still in high school in Illinois:

There's not a better team in the state!
We've beaten La Salle at any rate,
Pat slings the opponents around
Like a wind makes apples fall to the ground.
See how we won in the tournament!
Our team light up the firmament!

Streator is becoming popular it seems
Just because of its High School teams.
Our sky is blue, though clouds are dark,
Our basketball team is surely a shark;
We're champions! We're champions! Who says we are not?
We'll knock that fellow off the school block.

Only nine years after penning this charming boast, Clyde discovered Pluto, the first of many discoveries that marked his career. One of his most interesting occurred in 1936, when he detected what he called a "great stratum" of galaxies stretching from Alpheratz in the square of Pegasus to Algol in Perseus. More than 20 years would pass before astronomer George Abell cataloged the northern sky's great supercluster of galaxies. More recent observations have turned up a second supercluster, in the constellations of Ursa Major and Lynx, and the two groups may be related, connected gravitationally in some way. If that is true, Clyde's great stratum stretches from horizon to horizon — across half the sky.

I miss the passion of the man, the humor, and the ideas. He had a brilliant command of English that allowed him to pun at will. His famous crow jokes were especially bad: "Where do the crows go to drink? A crow bar, of course." Even this crowing had something of a scientific edge to it. While doing research for this book's first edition in the late 1980s, I uncovered Clyde's unconfirmed discovery of a nova in the constellation of — what else? — Corvus, the Crow. I first mentioned this find in the Appendix, which I wrote in 1990, but the story of Nova Corvi 1931 continues to this day. One of the star's rare outbursts was witnessed in 1991 by the International Ultraviolet Explorer satellite. It delighted Clyde to know that those few photons of starlight captured on his photographic plates would lead to an observation by an orbiting telescope. Then, in February 2005, I happened to get a "movie" of an outburst — they occur only for two hours about once a year — just as the star was rising in the east.

I miss Clyde's enthusiasm for life. When I'd accompany him on cross-country lecture tours, I could only marvel at the energy and drive he'd maintain for days on end. As he grew frail in his last years, Clyde spent more time in his favorite living-room chair, marking up his newspapers, magazines, and books with marginal notes.

On Friday morning, January 17, 1997, Clyde suddenly slumped in his wheelchair and was gone. Death, noted Patsy, his wife of 62 years, had come as fast as if a light had gone out. On the day of the memorial service, their modest home brimmed with the people who knew him as a brother, a husband, a father, a grandfather, a colleague, and a friend. But his favorite living-room chair was empty — no one dared sit there. There was warmth in the house that day, as we sat and remembered Clyde, all around his empty chair. Clyde's journey was over.

I miss the observing experiences we shared, particularly one in the spring of 1983. In my logbook he wrote, "Sure enjoyed looking at the many objects with your 16-inch telescope, especially Pluto, the Hercules globular cluster, and the Whirlpool Nebula, M51. I could see the spiral structure in M51. Also, M13 was superb! But poor little Pluto was so faint, unimportant looking. . . . Thank you for this observing session."

These words hint at his greatest achievement — and the seed of one of his greatest troubles. In his last years Clyde was dogged by the scientific debate about whether Pluto — his signature discovery — qualifies as a true planet. "If it quacks like a duck and looks like a duck, then it *is* a duck," he would say. But he worried that the import of his accomplishment would be reduced immensely if Pluto were to be regarded as an asteroid, one of hundreds of thousands, instead of one of a few planets.

In the years since his death, a new planetary paradigm has evolved. When Clyde discovered Pluto, he didn't realize that he had opened the door to the Kuiper Belt, of which Pluto is the most famous member. We didn't take our first tentative steps through that door until the discovery of a second Kuiper Belt object, 1992 QB_1, more than six decades later, and the hundreds of others that followed. As I write this preface a new body — designated 2003 UB_{313} and almost certainly larger than Pluto — has emerged from the darkness of the Kuiper Belt.

How delighted Clyde would have been to participate in a visit I had with Patsy on November 5, 2005. She had just heard from the discoverers of two new moons of Pluto, seen on images taken by the Hubble Space Telescope. Clyde's world, now a system of a planet and three moons, gets ever more exciting. And a spacecraft, dubbed New Horizons, has just begun the long interplanetary trek to meet up with Pluto and its moons in 2015.

Interestingly, Dad's story about the outer planets left its ending open. He had a hunch that there might be other major planets out there, and that Pluto would not be the last. Now astronomers imagine perhaps dozens of icy dwarf worlds like 2003 UB_{313} standing guard amid thousands of smaller comet-size bodies at the edge of our planetary system.

Clyde never looked at the sky in small doses. By discovering Pluto in 1930, he opened the door to what might be the largest observable group of bodies in our solar system. Six years later, his stratum of galaxies presaged the revelation that distant regions of the universe are filled with great superclusters. Moreover, he did all this alone, in a cold dome with no warm room and no computer, his hands attached to the control paddle of a telescope. Knowing Clyde, he wouldn't have wanted it any other way.

PREFACE

WHEN I first encountered Clyde Tombaugh at a symposium on the exploration of Mars in Denver in 1963, I was impressed with the modesty and good humor of a man who had gone through life with the distinction of having discovered a major planet. Like that of William Herschel, who discovered Uranus, Clyde Tombaugh's achievement in finding Pluto gave him both a tremendous start in life as well as a sense of noblesse oblige — what does one do after such an accomplishment?

From that brief meeting in 1963 to the start of a biography in 1984 required a little growth on my part. Over many nights of observing, and my own sense of discovery with some comets, I realized that one understands more about an episode if one knows something about the person most deeply involved with it. As I got to know Tombaugh, his modesty proved only part of a complex personality, and his humor turned out to be an incisive and mature way of enjoying his life.

Accordingly, this project is not intended to be a scientific biography or a scholarly work of astronomical history. It is an evaluation of the life of a man I have come to admire, and my purpose in writing it is to explore the feelings, the motivations, and the many joys that Clyde Tombaugh has experienced. My purpose is also to inspire. The story of the discoverer of Pluto is a good example of the type of experience that makes an astronomer's life exciting, varied, and full of change.

The basis of this biography is a long series of interviews of Tombaugh, his colleagues, friends, and family. I have not provided specific citations for every quotation from the interviews with Tombaugh, except for those statements I deem especially important or controversial.

The story does not rely exclusively on this type of source, however. Personal recall is necessarily subjective, so my research for the Lowell chapters included checking every one of his plate envelope notes, as well as whatever observing records I could find. Excerpts of a diary kept by Carl Lampland which described some of Tombaugh's activities around the discovery of Pluto in 1930 were kindly provided by his colleague Henry Giclas. The complex events involving how the discovery was reported and the providing of precise positions for orbit calculation were discussed with Brian Marsden, current director of the International Astronomical Union's Central Bureau for Astronomical Telegrams. Dr. Marsden's insights confirmed my own suspicions that the senior observatory staff did not handle that event very well.

For later chapters, what I learned from the interviews was checked wherever possible against original written materials such as project proposals. I particularly appreciate the evaluation by University of Arizona planetary scientist Brad Smith of the chapters dealing with astronomy at White Sands and the years at New Mexico State. Walter Haas, founder of the Association of Lunar and Planetary Observers and thoroughly familiar with many aspects of Tombaugh's life and times, read the manuscript and enriched it with his many suggestions.

Checking out some stories led to a trail of unexpected events. Tombaugh's mention of the discovery of a comet led me to find that the comet had never been reported. As a result, I went on a long chase for confirming images of the object. I also found that Tombaugh did not recall his discovery of a nova in 1931, and my subsequent investigation of old Harvard patrol plates led to evidence that the object has recurrent outbursts and is still active. This story is told in the appendix.

Other incidents led to journeys far from science. The final chapter offers an account of Tombaugh's acting as honorary coach at a varsity football game at New Mexico State. Although Tombaugh remembered the details, he did not recall when exactly the game took place or the name of the opposing team. For a biographer who knows nothing about football, finding the missing details turned out to be fun. The trail led from Tombaugh's honorary coach-of-the-year plaque, to a discussion with NMSU's Sports Information Department, which, it turned out, had an archival record of every play in every game!

This biography began in 1983, when Tombaugh absolutely refused my request to do one. Disturbed by inaccurate press reports about his work, Tombaugh hoped at the time to write his own life story. By the end of 1984, I had approached him a second time with a more solid proposal and a tentative schedule of interviews. This time he was more interested, and the interviews began in January 1985 at the American Astronomical Society meeting in Tucson.

By mid-1985, interviews were in full swing. Every month I would drive the four or five hours from Tucson, Arizona, to Las Cruces, New Mexico; after dinner Clyde and his wife, Patsy, and I would examine a period of his life in a session that would last at least until midnight and often much later. (After all, we are observers used to night schedules.) The following day the interview would continue. These trips were supplemented by several visits to Flagstaff. In the fall of 1987 I accompanied Clyde and Patsy on a major speaking trip to some northeastern states, and in 1988 I met him in other places. By this time, with stories already told and retold in response to different approaches to questioning, the monthly schedule slackened a bit. The writing phase in 1989 brought a return to frequent Las Cruces trips, so that Clyde and Patsy could check each chapter for accuracy while I investigated other sources.

The result is a book that I hope will be fun to read. I have tried to paint a picture of a person whose life is dominated by observing in many fields, by a passion for observing the worlds around him.

Acknowledgments

This project involved the kind assistance of many people. To Bradford Smith, Jed Durrenberger, G. Harry Stine, Joe Gold, Austin Vick, Joe Marlin, Gordon Solberg, Frank S. Hemingway, Scott Murrell, Herb Beebe, Alden Tombaugh, and Annette Tombaugh, I am indebted for the interviews they granted. Ellen Willoughby, Brian Marsden, Eugene Shoemaker, Jean Mueller, Fred Whipple, Henry Giclas, James B. Edson, Jim Scotti, Roy Stowell, Mark Coco, Edith Levy, and Ed Mannery provided useful material and helpful criticism.

Walter and Peggy Haas gave invaluable and constructive criticism of the manuscript, and Arthur A. Hoag, Peter Collins, Pete Manly, Steve Larson, and Lonny Baker provided useful suggestions. Richard Berry, Robert Burnham, and David Eicher were helpful with Chapter 7. I wish to thank Brian Skiff, Michael Magee, Jeanine Cockrell, Norman G. Thomas, Carolyn Shoemaker, and four anonymous reviewers, whose suggestions strengthened this book. Also, the Lowell Observatory kindly made available all of Tombaugh's original photographic plates and plate notes for my examination. I appreciate the work of the staff of the University of Arizona Press, particularly Barbara Beatty who was helpful during the project's formative stages and copyeditor Kim Vivier.

I especially thank Judy Stowell for her careful transcriptions of more than eighty hours of recorded interviews, Patsy Tombaugh for years of helping with interviews and manuscript suggestions, and finally Clyde Tombaugh, whose help and friendship I shall always cherish.

Clyde Tombaugh with cat Pluto, 1990.
PHOTO BY THE AUTHOR.

*Clyde Tombaugh,
summer 1907.*

The Tombaugh family: Muron, Clyde, Esther, and Adella.

The Burdett High School track team, 1925.
Clyde Tombaugh is in the front row, second from left.

Robert Tombaugh looks through
Clyde's 5-inch single-lens refrac-
tor, circa 1935. Clyde reground the
lens into a secondary mirror for a
16-inch telescope around 1970.

The 9-inch in 1989.

On Mars Hill, circa 1936. From left: unknown, Art Adel, Irene Edson, Catherine Adel, James Edson, Patsy Edson Tombaugh, unknown, Carl O. Lampland, Mrs. Lampland, Lil Edson, Emma (Mrs. V. M.) Slipher, V. M. Slipher; front row: Bernice and Henry Giclas, and Henry Giclas, Jr.

Clyde and Patsy Tombaugh, circa 1935–36.

Annette and Alden, Los Angeles, circa 1945.

Clyde Tombaugh and the 16-inch, Las Cruces, 1989.

*Clyde Tombaugh and Susan Arm-
strong at the Beaux Arts ball,
circa 1974. Dressed as an Italian nobleman, Clyde won first
place; Sue won second place for her
Monopoly board costume.*

One of two posters used for New Mexico State University's Clyde Tombaugh Scholars Fund. Against a Pluto-Charon background, Gini Thatcher's artwork depicts, counterclockwise from top left, *Tombaugh and his 9-inch reflector, Lowell's 13-inch dome, the 13-inch A. Lawrence Lowell Astrograph, the blink comparator, and the two Pluto discovery plates. For the discovery of Pluto's moon Charon, the poster shows Tombaugh with James Christy, the 61-inch reflector at the U.S. Naval Observatory in Flagstaff, James Christy discovering Charon, and the discovery image of Charon to the upper right of Pluto.*

Detail from the 13-inch Lawrence Lowell Astrograph plate of January 10, 1931. "Comet 1931 AN," discovered by Tombaugh, is in the center. Courtesy of Lowell Observatory.

Circular No. 4983

Central Bureau for Astronomical Telegrams
INTERNATIONAL ASTRONOMICAL UNION
Postal Address: Central Bureau for Astronomical Telegrams
Smithsonian Astrophysical Observatory, Cambridge, MA 02138, U.S.A.
Telephone 617-495-7244/7440/7444 (for emergency use only)
TWX 710-320-6842 ASTROGRAM CAM   EASYLINK 62794505
MARSDEN or GREEN@CFA BITNET   MARSDEN or GREEN@CFAPS2.SPAN

CATACLYSMIC VARIABLE IN CORVUS

D. H. Levy, Tucson, AZ, communicates the discovery by C. W. Tombaugh of a cataclysmic variable in Corvus on a plate taken 1931 Mar. 23 with the 0.33-m A. Lawrence Lowell astrograph at Lowell Observatory, when the star's magnitude was ~ 12. A search by Levy through 260 plates in Harvard College Observatory's archives revealed 9 other maxima on the following dates: 1932 Mar. 2, 1940 Feb. 12, 1941 Feb. 5, 1952 Apr. 21, 1971 Apr. 20, 1983 Feb. 22, 1985 Mar. 15, 1987 Mar. 6, and 1988 May 21. Levy reports another outburst on 1990 Mar. 23.270 UT, with the star at m_v = 13.6. Precise positions measured by B. Skiff, Lowell Observatory (equinox 1950.0): discovery plate, $\alpha = 12^h17^m48^s38$, $\delta = -18°10'27''4$; Palomar Sky Survey O exposure, 1954 Mar. 7, $\alpha = 12^h17^m48^s64$, $\delta = -18°10'22''7$ (estimated blue mag ~ 17-18, red mag ~ 19).

COMET AUSTIN (1989c₁)

D. G. Schleicher and D. J. Osip, Lowell Observatory; and P. V. Birch, Perth Observatory, report: "We have obtained gas production rates based on aperture photometry obtained on 6 nights between 1989 Dec. 19 and 1990 Mar. 7 using the Lowell-Perth 0.61-m telescope, and on Mar. 14 and 15 using the Lowell 1.07-m telescope. For mid-March, log $Q(C_2)$ = 26.6, log $Q(C_3)$ = 25.4, and log $Q(CN)$ = 26.7 (i.e., the relative abundances are basically normal). $Q(C_2)$ varies approximately as $r^{2.0}$ over the total observational interval; however, during the first month the increase was much steeper than the mean, varying as r^4, while from mid-January to early March it varied as r^1. The March observations imply that a higher rate of increase may have resumed. It is unclear how much, if any, of these changes in slopes are due to short-term variability. The increase in the dust production has been extremely shallow, with log $(Af\rho)$ = 3.2 in mid-March (A being the albedo of the grains, f the filling factor of the grains, and ρ the radius of the field of view; cf. A'Hearn et al. 1984, A.J. **89**, 579), and showing variation as r^n with $n < 1.0$ since December. These data imply a current gas-to-dust ratio approximately 3 times higher than was observed in P/Halley at a comparable heliocentric distance."

Total visual magnitude estimates (B = binoculars): Mar. 19.11 UT, 6.1 (A. Hale, Las Cruces, NM, 10×50 B); 22.01, 6.0 (J. E. Bortle, Stormville, NY, 15×80 B); 23.13, 5.9 (C. S. Morris, Whitaker Peak, CA, 20×80 B).

Circular No. 4983 of the International Astronomical Union.

1990 March 23   *Daniel W. E. Green*

Chapter 1

LOOKING INTO
CHAPMAN'S HOMER

T HAT'S IT!" The words came out more as a thought than as something spoken. There it was, a starry speck on each of two photographic plates, a speck that had moved in relative position from one plate to the other by the amount he had hoped for. Ten months earlier Clyde William Tombaugh had begun his search for Planet X, and already, it seemed, he had found it!

February 18, 1930

On the cloudy morning of Monday, February 18, 1930, Tombaugh woke around seven o'clock in his bed on the second floor of Lowell Observatory's main building. Hired to take photographic glass plates at the telescope and to "blink" those plates for the unknown planet, Tombaugh divided his time between these two very different tasks. Since the bright last-quarter Moon had prevented any observing the night before, blinking for planet suspects was the agenda for this day. Tombaugh's thoughts were on two photographic plates he had taken on January 23 and 29. *Blinking* is a method of examining portions of two photographic plates using an optical system that alternates quickly between one and the other. It is one of the simplest jobs an astronomer can do, but one that requires a degree of concentration and insight few people can achieve. Although Tombaugh had been hired only for planet-search photography, for the past few months he had been given complete responsibility for the blinking as well; the entire trans-Neptunian planet search was his.

A light breakfast at a downtown café started the day. During his morning trip down Mars Hill to Flagstaff, Tombaugh stopped by the post office to get the mail and then returned up the wooded hill to the observatory. Lined with tall ponderosa pines, this curvy road spends its winters in a green-white tranquility, a mountain scene vastly different from the family home in Kansas he had left a year ago to take up his new position at Lowell.

By nine in the morning Tombaugh was inside the blink comparator room in the main building. Three days earlier he had mounted these two plates in Gemini for blinking, one in each position on the comparator, and carefully aligned them. It was vital that the plates be perfectly matched in four ways: in length of exposure and sky conditions, so that the relative star brightnesses were the same; in position, so that one field of view was identical to the other; in time of exposure, so that one plate would not be affected by refraction more than the other; and in the way they were set up on the comparator—the centers had to be matched because the image distortion at the edges had to be identical. Both the January 23 plate, No. 165, and the one for January 29, No. 171, were the results of one-hour exposures with the 13-inch A. Lawrence Lowell telescope. Later Tombaugh inscribed on the envelope for 171 that the exposure had lasted from 9:13 to 10:13 and that the interval between the exposures was six days, a "Fairly good pair to blink." Noting also that it was a "Cramer's Hi Speed Plate," he added that it had been processed using Rodinal developer. From February 15 to 17 he could "devote only part-time to blinking," but now he was ready for a full day.[1]

Using a microscope eyepiece and the blink comparator, Tombaugh examined the two identical areas of sky for any changes. During a search three kinds of changes would invariably catch his attention. One change is an alternating brightening or fading of a star; since many stars in the sky vary in brightness, such a change, though interesting, would not normally be a cause for excitement. Another change occurs when a star is displaced a short distance from one image to another, so that as the two images are examined in rapid succession, the object appears to move back and forth. A third change is the sudden appearance of an object that, invisible on the earlier plate, bursts into view on the second. A rapidly brightening star or nova would produce this effect, as would any moving object that had traveled some distance during the time between the two exposures.

The monotonous *clack, clack, clack* of the electromagnet as the optics shifted from one plate to another and back was tiresome and draining, and after a half hour he took a short break. After another half hour he stood up and walked away for perhaps fifteen minutes. "I knew I dare not overdo it or my attention would lapse and I could miss something. This haunted me all the time." On this typical blinking day Tombaugh planned a total session of about nine hours at the comparator. He had set a blinking rate of three changes per second from one plate to another. If the rate were much faster—say, ten times per second—he could not notice any changes. A slower rate of one or two changes per second would have added significantly to the search time.

During a long break Tombaugh might plan a night's observing, develop plates from a previous night, read a journal, throw logs into the furnace—anything to get his mind away from the mental concentration that blinking required. At noon February 18, lunch followed more than two hours of blinking. Perhaps he thought that without any observing tonight, he might get as much as four more hours of blinking done today. It would be a slow task, for not even a quarter of the two plates had been covered since he began.

Since Tombaugh's tiny living quarters were not well stocked with food, he took the ten-minute drive downtown to the Black Cat Café for almost every meal. One of the first buildings Tombaugh noticed after he arrived in Flagstaff, the place was right across the street from the railway station. He ate there so often that the waitress didn't even have to ask him for his order. If the owner wasn't cooking or cleaning up, he might come over to chat about what Tombaugh was up to. On snowy days the road might be so blocked with snow that the observatory was locked in, although a horse-drawn sled was available for quick snowy trips for supplies. After this monotonous morning of blinking, Tombaugh was pleased that the road was open.

By one o'clock Tombaugh had returned to the blinking room. Over the soft background of clicking sounds as the comparator moved its glance from one plate to another, the search progressed in careful steps from the edge to the center of the plate. The eyepiece was comfortable, set up at a convenient angle so that Tombaugh's eyes were quite relaxed throughout the long session. It was his mind that needed a rest from the intense concentration, so just after one-thirty he took another short break, and a longer one in mid-afternoon. Now he might have read an astronomical journal paper or something to give his mind a rest. When E. C. Slipher was at the observatory, the two often got to talking about Mars or some subject that could keep Tombaugh from the comparator for an hour or more. But on this winter afternoon Lowell's resident planetary expert was away at the Arizona legislature in Phoenix.

As the search progressed closer to Delta Geminorum, Tombaugh ruminated on how nice it was that he had decided to skip, for the time being, the thickest part of the Milky Way to the west. On these plates the stars were about as dense as he would want to search through. Since the plates were far too large to be examined at once, they were divided into eight panels, and he examined each panel in a series of four or five strips. At the end of a panel he would place a mark and then lower and reset the plates and begin the next panel.

Tombaugh was looking for an object moving at a specific rate and direction. After searching thousands of stars in a pattern moving across the plate, he returned to the starting point to begin a second strip. After several of these horizontal strips, Tombaugh marked his place and turned the plates around. The object he was looking for would now be moving in the opposite direction. "I had to remember that retrograde was now the other way, just like a football player has to retrain his mind after every quarter, that his goal has changed." Keeping in mind these new "goalposts," Tombaugh continued his eastward trek. Finally, around 3:50, he encountered bright Delta Geminorum, the star he had guided on during the exposure. Shining prominently, it monopolized his field of view. Outside, the cloudy sky was already beginning to darken.

It was now four o'clock in the afternoon. One-quarter of the way through the pair, he blinked on a bit further. Although the stars on these plates were thick, about a thousand per square degree, the only differences between the two plates in this field of view were faint inconsistencies that Tombaugh knew were just defects in the emulsion. Delta Geminorum moved away as Tombaugh moved one field away, then two, and three. Then, almost as an afterthought, an object appeared, then disappeared, on one of the plates.

"I saw a little image popping in and out, looked to one side, and saw another one doing the same thing." They seemed to be out of phase with one another, so that when one appeared the other disappeared. The objects were about fifteenth magnitude. Had the two plates been taken only two or three days apart, the object would have appeared to shift position instantly from one plate to the other as the blink eyepiece moved its attention from one to the other. Because the interval was a relatively long six days, Tombaugh saw the images appear and disappear separately. They looked suspicious, but the comparator was blinking quickly. Suppressing his excitement, he turned off the automatic blinking and thought. Could they be simple defects in the emulsion that had been applied to the plate—artificial, dark, starlike spots that happened to be placed at the right places to fool a careless searcher?

Could the images be a slowly moving asteroid? Because the plates had been taken at an area in the sky opposite the Sun, known as the opposition region, asteroids in the main belt between Mars and Jupiter would be moving in the same direction and at about the same rapid rate. Because of the effect of the Earth's own orbit around the Sun, asteroids appear to slow when they are far from the opposition region. Grabbing a plastic millimeter ruler, Tombaugh measured the shift in position more accurately; it was 3.5 millimeters from one plate to the other. There was now no question that if the object were real, it had to be beyond the orbit of Neptune.

With the comparator's motor stopped, Tombaugh then inspected the plates manually to confirm that they showed objects at two different locations. "I took note of the dates of the plates. The image on the earlier plate was to the east, the one on the later was to the west; it was retrograding."

Only a few months after its discovery, the new object would be almost lost in the glare of Delta Geminorum. Would Tombaugh have missed finding it in that case? "I was well aware of Murphy's Law, although I didn't know it by that name at the time. I had taken several plates of Delta Geminorum and I had casualty after casualty; I should have realized that the little rascal was trying to thwart me. Those plates were jinxed." Now thinking that one of the problem plates had been exposed on January 21, 1930, Tombaugh moved to get it.

Could these be images of two variable stars? The answer would be in this January 21 plate, which, Tombaugh remembered, had been exposed under the worst seeing conditions he had ever experienced. Just after the exposure began, a strong wind had rushed up from the north-east and suddenly the star he was guiding on swelled virtually beyond recognition. Since he had gone to the trouble of setting up and loading the plate, he thought he might as well complete the exposure before shutting down for the night.

It was this awful plate that Tombaugh now summoned. "The thought occurred to me I'd better get out that third plate and see if there was an image about 1 millimeter east of the position on the 23rd plate; if there was, that would pretty well cinch it." Although the images were "soft"—larger star images with poorly defined edges, rather than sharp points of light—the new object was there, shifted precisely the amount Tombaugh hoped for. "I was walking on the ceiling. I was now 100% sure."

Across the hall, astronomer Carl Lampland was working in his office, barely paying attention to the clicking of the comparator as it automatically moved the shutter from plate to plate. But for the past half hour the clicking had stopped. Tombaugh was still in the room. Thinking that the young Tombaugh must have stumbled onto something, Lampland waited and wondered.

About thirty minutes had passed. Just one more thing: mounted on the 13-inch camera was a smaller 5-inch camera, and concurrently with every plate taken with the large instrument, a 5-inch camera plate had been exposed as well. Could the faint images exist on these smaller plates? Moving quickly now, Tombaugh removed the smaller plates from their brown envelopes and examined them. Because the two images were very faint, it took some time to locate them, but they were definitely there. "I can hardly describe to you how intense was the thrill I felt. I was looking through my hand magnifier identifying the configuration of the stars. I could hardly see them; my hand was shaking; I was literally shaking with excitement."[2]

Tombaugh put the small plates away and replaced and aligned the January 23 and January 29 plates. Three-quarters of an hour had now passed. As he said later, for all that time "I was the only person on Earth who knew where Planet X was!"

It was time to announce.

Across the hall, Lampland was in his office. "I heard the clicking sound and then a long silence. I had suspected that you had found something," Lampland said a few days later. Tombaugh adds, "Then I realized that poor man must have been sitting at his desk that three-quarters of an hour just dying of suspense waiting for me to invite him in!" Tombaugh could have been resetting the plates, but that takes only a few minutes. He could have been taking a break, but he hadn't left the room. Although he had done considerable manual shifting from one plate to another, the clicking sound thus produced was much softer. Lampland had become suspicious.

Tombaugh crossed the hall to Lampland's office, looked in, and said, "I think we've found Planet X." The older astronomer darted across the hall and looked through the comparator eyepiece, studying the images and Tombaugh's explanation of the shift. Later he noted in his diary that Tombaugh "found an interesting object today and it is to be hoped it may be our long sought planet."[3]

While Lampland was examining the plates, Tombaugh left the room again. He walked down the long corridor, past the secretary's room, and stopped just outside V. M. Slipher's large office. "I wanted to appear somewhat nonchalant, not too emotional about it." Examining some papers, the director looked up as Tombaugh entered.

"Dr. Slipher," Tombaugh said, "I have found your Planet X."

The director rose as though something had propelled him from his chair. He stared as Tombaugh added, "I'll show you the evidence," and then hurried to the comparator room. Those must have been the longest ten seconds of Slipher's life. He had been involved now with Planet X for twenty-five frustrating years. Could this really be it? Tombaugh had never reported a false planet suspect. Why hadn't Slipher found it himself? Would this young man get the credit if the discovery were confirmed? Quite likely, the seeds for the observatory's treatment of Clyde Tombaugh during the next fifteen years were planted during the brisk walk to the comparator room.

By now Lampland was convinced. "It looks pretty good," he said, as Slipher rushed in. During the next hour the three men looked at the images and considered some of the implications of this discovery. With only one chair in the room, all three remained standing. Slipher had two important thoughts: first was a suggestion that the Delta Geminorum region be rephotographed immediately, but as they looked out the window they suspected that the solid deck of clouds would not break in time for that evening. The Moon would not rise until midnight, offering some dark sky if the clouds dispersed. The second thought: no one outside the observatory staff, absolutely no one, was to know about this development. E. C. Slipher would be informed when he returned from Phoenix in a few days.

With the sky over Flagstaff now completely dark, the conversation ended around six o'clock. Not even realizing that the three had been standing up all that time, the group left the room, Lampland and Slipher going to their homes and Tombaugh heading toward the downtown café. As he usually did, he picked up the observatory mail and went to dinner. He did not recall what he ate: "you could have probably fed me bear meat and I wouldn't have known the difference." Unable to concentrate on his dinner, Tombaugh's head was filled with three thoughts: that Planet X had been found, that he had found it, and that he could not tell anyone, not even his parents!

Tombaugh was thorough, careful to follow his superior's instructions, careful to record all that he did. By the middle of June, he had written down the details of his examination of plate 171, which of course was only one-fourth completed on February 18. On the back of the plate's brown cover, his neatly scripted notes read:

> Examination began about Feb. 15, 1930; but I could devote only part time to blinking. The thickness of stars slowed up the work considerably. At the end of the day of Feb. 18, 1930, about ¼th of the pair was examined. On February 18, 1930 at 4:00 PM (Mt. Standard Civil Time)—planet X [Pluto] was discovered—using the comparator. The following ¾ hour was spent investigating the amount of shift or apparent motion of Pluto, also its direction, and reality of images. The last mentioned was settled by the identification of the images in their respective positions on the 5-inch Cogshall Camera plates, which were made simultaneously with the 13-inch telescope discovery plates [No. 165 of Jan. 23 and No. 171 of Jan. 29]. A third plate of Jan. 21 also showed the planet in correct position with respect to the other dates. At 4:45 PM the discovery was made known to Dr. C. O. Lampland first, and then to Dr. V. M. Slipher and then to Mr. E. C. Slipher. I shall never forget the intense interest that the object caused among the rest of the staff. No further examination was done for a considerable period—as Dr. Lampland used the comparator for measurement of positions on the 13-inch plates, and also for identification of Pluto images on plates that he made with the 42-inch reflector on

following nights for several months. Examination of this pair of plates was resumed about May 26, 1930, but could devote only a small part of each day, and examination of the entire area of the plate was not completed until June 9, 1930. The Ottawa object was watched for the remaining ¾ths of the plate, but nothing else of a transneptunian nature was found.

—C. W. TOMBAUGH

June 9, 1930.

Notes of Examination:—1. 8 asteroids "A". 2. 7 variable stars "V". (+ 3 or 4 others not marked on this plate—see plate 165 for their positions.) 3. 12 temporary objects (on one plate only) which may be variable stars, novae, or defects. Two different suspicious objects (transneptunian) of 17th mags. were found to be a variable star in both cases, from checks on other plates—on one plate a variable and a defect on the other. 4. No comets. 5. Planet "X" (Pluto) at last found!!! [P] [4]

It was cold, cloudy, and late on February 18. Thoughts about these events filled his mind during his dinner, which could not have lasted long. He left the café and looked skyward again; the thick clouds covered the entire sky. He was so tense he could not bear the thought of returning to the observatory to do more blinking; in fact, he thought with a smile that his blinking days were over. Although there was snow on the ground, at least it wasn't snowing. After all, for the next plate he would not need a completely clear sky, just clearing in the Gemini region for about an hour. It might clear later; perhaps in the meantime he would walk down the street.

Just a few blocks away was the Orpheum Theater, where *The Virginian* was making its first run. Starring Gary Cooper, this classic western film was a story of love and adventure, with a slow build-up toward a gunfight at the end. As it was somewhat slow-paced at the beginning, Tombaugh did not concentrate fully on the story and did not notice the plot twist toward the final street battle until it was too late. Gunfire, ricocheting bullets, and characters falling—it all happened in an instant, and Tombaugh practically jumped out of his skin! Minutes later the film was over.

Tense and excited, Tombaugh approached the theater door hoping that at least the clouds had thinned a bit, hoping for a sign that he might still get a picture of Delta Geminorum. No such luck; the clouds were thick as ever. Driving back in the observatory's Ford, he put the vehicle in the garage, went to the darkroom, and loaded a plate into one of the plate holders, just in case.

These plates are very large by today's standards, 14 by 17 inches. Carefully, he loaded the plate into its plate holder, curved gently to match perfectly the focal surface of the camera. Bending the glass just $3/16$ inch was absolutely necessary; if that were not done, the edge of the plate would be out of focus and useless for finding new objects. Every time Tombaugh loaded a plate, he recalled an awful night a few months earlier when at midexposure, the dome reverberated with a huge crash. It was loud enough to be the telescope's main Alvan Clark lens, and had that disaster occurred, the whole program might as well have been over. Going fearfully to the front of the telescope, Tombaugh looked at the lens, which fortunately was still intact. Only the film plate had broken. The following day he reported the incident to Slipher, who replied that curving the plates would break one every now and then and that there was nothing he could do about it. Two or three more of these shattering booms were enough to send Tombaugh to a design session—for his own sanity, if not to save five dollars per broken plate. The plates, he reasoned, were held in place by adjustable screws that were gradually tightened until the proper curvature was reached. By carefully loosening the corner screws to allow the plate to spread, and then locking them, Tombaugh felt he could avoid the problem. It worked, saving plates and nerves. Tonight, as he mounted this plate with his new procedure, he reflected on that earlier success he had had as a Lowell employee. Carefully loosening the screws on the plate, he then stored it in the darkroom, all ready to go should the sky clear. The astronomers must be pretty pleased with him now, he thought.

At the moment, there was nothing to do but wait. Walking around the observatory, he spent some time looking at Lowell's books in the reading room. Would that their original owner could be around to share this night! Tombaugh recalled the day last spring that Director V. M. Slipher had given him this new work and how appalled he was at the added obligation. Now he felt pleased that his reasoning out the mechanics of the problem had led directly to the discovery; a half-century later he found that Lowell himself had come up with the same suggestion to search near the opposition point, but that somehow the observatory staff had long forgotten it and had never told Tombaugh.

Always a dark and quiet place, the observatory must have been especially creaky that night. Its reading room contained two large pendulum clocks, one set for local time and the other, with a slightly shorter pendulum, for sidereal time; their tickings were not synchronized. The room was a repository of the knowledge Lowell had gained, knowledge that had led to his hopes to extend the solar system. Tombaugh looked at all those Mars books that had led to Lowell's early enthusiasm and later distress, and he reflected on Lowell's hope that the finding of a fresh new planet would bring stature back to the observatory.

Lowell was right. For hundreds of years discovery has brought fame and respect to those with the stamina to follow a program that leads to one, or the luck to stumble onto one. There was William Herschel's discovery of Uranus, an event that had taken place on March 13, 1781, close to a century and a half earlier. Like Tombaugh, Herschel had begun his career as an amateur astronomer, and also like Tombaugh, he enjoyed building his own telescopes and observing with them.

Repeatedly, Tombaugh walked outside to gaze at the thick clouds. The sky looked dismal, with not the slightest sign of clearing. Back inside, then outside, Tombaugh repeated his pattern until well past midnight, when the last-quarter Moon rose. Even if the sky were to clear now, the Moon's bright light would have rendered conditions unsuitable for a photograph. Disappointed but still excited, Tombaugh gave up at around 1:30 AM.

Usually able to drift off to sleep quickly, tonight Tombaugh lay awake in bed, his mind filled with thoughts like these of a poem written by John Keats in October 1816. The sonnet "On First Looking into Chapman's Homer" lives in the spirit of poets and observers. On the night of February 18 Tombaugh was the only living person on Earth for whom this poem had literal meaning:

> Much have I travell'd in the realms of gold,
> And many goodly states and kingdoms seen;
> Round many western islands have I been
> Which bards in fealty to Apollo hold.
> Oft of one wide expanse had I been told
> That deep-brow'd Homer ruled as his demesne;
> Yet did I never breathe its pure serene

Till I heard Chapman speak out loud and bold:
Then felt I like some watcher of the skies
When a new planet swims into his ken;
Or like stout Cortez when with eagle eyes
He star'd at the Pacific—and all his men
Look'd at each other with a wild surmise—
Silent, upon a peak in Darien.

Exhaustion had set in. For Clyde Tombaugh, the day had ended.

Chapter 2

THE EDUCATED ONES

Streator, Illinois

IN MY STARS I am above thee; but be not afraid of greatness. Some are born great, some achieve greatness, and some have greatness thrust upon 'em."[1] This Shakespearean line comes from a letter read by Malvolio, steward to the countess Olivia in *Twelfth Night*. For those who achieve greatness, what happens in early life can set a pattern; for Tombaugh, a combination of three factors worked in his favor. First, he was a firstborn whose siblings say acted like a second father for them; second, his father worked hard to instill in his family the values of hard work and doing things right the first time; third, his uncle had a serious interest in astronomy.

Learning was emphasized over several generations of Tombaughs. Jacob Tombaugh, Clyde's grandfather, was a Pennsylvania schoolteacher who had completed college, a man of such learning that he was exempted from serving in the Civil War. As a neighbor once recalled, "Tombaugh, oh yes, the Tombaughs were the educated ones."[2] Land in northern Illinois was rich and good for farming. Corn grew well there most years, and in the late 1800s Jacob Tombaugh moved out from Pittsburgh to homestead on land near Streator and married Sally Ostrander. Born in 1880, Muron Tombaugh was one of their four children.

Muron was a highly intelligent and motivated person, so well inclined mechanically that he had for a while pursued mechanical engineering at the University of Illinois. Although he did not finish, he

13

did continue his education informally through the International Correspondence School, a respected institution that provided well-written books and materials for people who could not attend a college. Muron never finished college, partly because of an illness and partly because, in those distant days, "he had the stigma of being a country boy."[3] "He quit college when he should have stayed," Clyde adds forcefully. "He was the youngest of his family, and he was a little on the spoiled side, and he didn't have the perseverance I had."

In 1905 Muron married Adella Chritton, a somewhat reserved person who had completed the eighth grade and who had a hardworking family background. On February 4, 1906, Clyde William Tombaugh, the first of six children, was born on a nearby Streator farm, in a solar system with eight known planets. Percival Lowell, convinced that a ninth existed somewhere in Libra, was one year into an intensive search. Clyde's sister Esther was born two and a half years later, his brother Roy was born in 1912, Charles in 1914, Robert in 1923, and Anita Rachel in 1929 as Clyde was traveling to Lowell Observatory in Flagstaff.

Clyde and the other children never felt that country-boy stigma that had bothered their father, partly because he wouldn't let them. Always energetic, Muron made sure his workshop was well equipped, and even though there were no electric tools, just hand-operated ones, he encouraged his children to use the tools and learn from them. "My Dad used to complain about our breaking drill bits," says Roy, "but he never kept us out of the workshop. I think some of those threshing days in Illinois had a lot to do with Clyde's getting into a good work habit." The ethic was not hard work but the fun of hard work; the sixteen-hour days at the height of threshing were long but fulfilling, days to look forward to. Muron had taught that the second try wasn't good enough; they had to do it right the first time. "We didn't talk about second chances," Roy said about their father. "One mistake was too many. I think that made a difference. How many hundreds of times could Clyde have missed that planet?"[4]

Clyde's grade-school years were happy ones, blessed especially with an enthusiastic fifth-grade teacher, Susie Szabo, who encouraged Tombaugh to pursue his interests in geography and history. The schoolhouse had two rooms, with three younger grades in one and five older grades in the other. Clyde was especially interested in geography. Miss

Szabo taught him to draw buildings, landscapes, and maps. As his work got better, it won prizes in the county fair. But Tombaugh's geographical interest prospered most on the farm, where he dammed the creek that traversed the pasture to create a pond with islands and other features that were excellent for ice-skating and miniature sailing. The dam was so successful that during threshing season, water haulers would come to collect water there for their steam engines.[5]

Tombaugh also enjoyed digging holes "to find out what was beneath." To solve a problem with a deeply laid pipe, over a two-week span he dug a hole twenty feet deep! The boy's curiosity extended way beyond the farm. What would landscapes look like, Tombaugh ruminated one day during a sixth-grade geography class, if he were to go to Mars? At first just a young boy's daydream, the study of the other planets' geographies dominated his attention in later years.

High school opened on a sour note. For two months of his freshman year he was out of school with a critically serious attack of whooping cough. Not able to hold food down, he almost starved. "I nearly dropped out of school for good, I was so discouraged. Two months of utter misery. And would 'whoop' at times for years afterward. It's a nasty disease, not to be taken lightly as some people think."

Finally over it, Tombaugh returned to school and eventually grew agile from bicycling seven miles every morning across gravel roads to Streator, and another seven miles back home every afternoon. In heavy rains the road would become wet mud. During his first high-school winter the roads were usually covered with snow and ice, making cycling treacherous. To ease the problem of long-distance travel, Clyde's paternal grandmother opened her Streator house for the Tombaughs and their two "double" cousins. (Lee Tombaugh, Muron's brother, had married Adella Chritton's sister.) Life in the Streator home that winter endeared Clyde to his cousins, and they became his closest friends.

With the return of spring Tombaugh resumed his fourteen-mile bicycle rides. With all this cycling, he developed powerful legs, serving him well in high-school sports like high jumping, pole vaulting, and baseball, all of which he excelled at later in Kansas. He also devised a football field in part of the pasture on the farm for Sunday afternoon touch football matches, and he built a tennis court behind the house.[6]

Although Tombaugh loved pole vaulting, the sport almost killed him around 1924. Clyde somehow did not notice that one of the poles he had carefully smoothed was cross-grained and fragile. One afternoon he ran hard with it and leaped up. Near the height of his vault the pole broke. Clyde hit the ground, winded and in pain. Roy rushed over to him: "He was just gasping for breath. If he had fallen on the spear of that broken pole, it would have been a disaster."[7] Never giving up, Clyde later used a replacement pole made of bamboo wood that was far safer.

THRESHING

The period from 1917 to about 1920 was difficult for farmers, since most of the young men were serving with the armed forces. Barely eleven years old, Clyde learned to plant corn when the weather became warm enough in May and to cultivate it during the hot days a month later. July was spent cutting oats and wheat, and the Augusts of his early teenage years were devoted entirely to the thresher.

The corn and wheat crops were what the success of the Streator farm was all about, and although Clyde enjoyed cultivating the corn, he dreaded having to pick it by hand during harvest season in late fall and early winter. By any stretch of imagination, the harvest must have been miserable. On cold, often frosty, weekend mornings Clyde would drive the horses as they pulled the wagon along, at the same time picking the corn and throwing it in the wagon. The mud was deep, and on rainy days he would be soaking wet after hours of labor. The invention of the work-saving corn huskers was years away.

Threshing, in contrast, was exciting and fun, a party whose center of attention was a large and noisy machine powered by a steam engine. On the Tombaugh farm about twenty-five men, organized into a neighborhood company Clyde's father ran, began around seven each morning, just after the dew evaporated, since dry straw threshes more efficiently. Pitchers would throw the bundles onto racks to be loaded into the thresher. Knives would cut the twine, and the straw would go down to a revolving cylinder whose teeth would tear the grain from the straw. Finally, a huge fan would force the straw out through a wind stacker. "The fellow behind the machine would open the door where the straw was going out through the cylinder," Tombaugh recalls. "Man, that was vicious-looking!" Clyde would watch the straw pour out through the cylinder as the steam engine drove the thresher at some eighteen hundred revolutions per minute.

In later threshing done by the Tombaughs in Kansas, the wagons would catch the grain to be taken to the bins or the elevator. Just before they stopped for lunch, the men would put the pitchforks, handle-first, into the haystacks to keep grasshoppers from attacking the wood. Sometimes the insects would come in such numbers that they would gnaw on the handles, making them hard on the hands. Meanwhile, the women would prepare a huge meal for the workers. The work would then continue till about six o'clock in the evening.

The steam engine made a loud rhythmic noise that Tombaugh enjoyed as he unloaded bundles of grain into the thresher. It was musical: "If I had to wait my turn I'd go sit on the steam engine with my dad. I learned much about steam engines; it's a love I've never forgotten." So much did the engines appeal to Tombaugh that for a time he envisioned a career beginning as a railroad fireman and going on to become an engineer. He liked being with his father during those times, and they had such fun together that someone new to the threshing gang insisted that they were brothers, not father and son.[8]

Clyde's father was the heart and soul of the neighborhood threshing company that served the nearby farms. Manager, engineer, and organizer, Muron Tombaugh worked day and night in a project he obviously enjoyed. During threshing season he would be up before four in the morning to fire up the steam engine so that all was ready to go when the team arrived at seven. After the long day's teamwork was over, Muron would cool the fires and, by the light of a kerosene lamp, adjust parts on the engine.

READING AND TELESCOPES

Early in high school Tombaugh took a course in general science that included both physical and biological studies. As a freshman he also developed an interest in ancient history. He read voraciously, continuing his daytime reading into the night with an old kerosene lamp. At first he concentrated on two books, the Bible and an encyclopedia. He read the entire Bible, a considerable feat for a young person who read it more out of sheer interest than from deep religious feelings. Eventually, he studied his father's books on trigonometry and physics and even learned a little Latin and Greek.[9] Not all Tombaugh's leisure hours were spent reading, just most of them. He carried this enthusiasm throughout his life, devouring books, underlining passages, and writing marginal notes.

One day his parents came home from the Streator Library with a book that had a color portrayal of Mars. "I was just absolutely fascinated with that. I thought, I've just got to see these things. I want a powerful telescope some day." Roy could never recall a time when his brother did not have an interest in the stars, and it appears that Tombaugh's interest in astronomy stretches back to early childhood. One Saturday morning in August 1918 while on a brief trip, twelve-year-old Clyde and his parents drove by Lake Geneva in northern Wisconsin. At midmorning they arrived at Yerkes Observatory and were greeted by a sign that instructed visitors that the observatory would not open until two that afternoon. The Tombaughs were eager to see the 40-inch refractor, one of the world's largest telescopes, but because of wartime gasoline restrictions, they had to be back to Chicago by the end of the day. Muron rang the doorbell and explained the situation. "My son is very interested in astronomy," he added. "Is it possible for him to see the large telescope briefly?" Of course not—the answer was disappointing but expected, although the observatory employee did spend a few minutes answering some of Clyde's questions about the size and type of telescope. It was a refractor with an objective lens over three feet in diameter that focused light at an eyepiece at the other end of the long tube.

The family continued on its way, stopping by the shore of Lake Michigan to watch the waves. What a view! It was Clyde's first encounter with large waves, and he sat fascinated for almost an hour as they washed in. A year later the family took another vacation, by train to Colorado. The trip included a visit to the summit of Pike's Peak, Clyde's first experience in the western cordilleran mountains that would later become such a rich part of his life.[10]

Every summer Muron's family visited the Lee Tombaughs, alternating at the home of one, then the other. Only nine miles away, Lee's farm provided a good observing site during these weekend visits. In 1918 Clyde first looked through a small telescope that his uncle had owned for more than twenty years. Lee was an amateur astronomer whose small telescope and one astronomy book were all Clyde needed to get started. The telescope was a 3-inch-diameter nonachromatic refractor, not unlike the telescope Galileo had used three hundred years earlier. Nonachromatic lenses do not correct for the different focal lengths of colors of the spectrum, so that the images one sees are fringed with unwanted color. This telescope had a power of only thirty-six, a little too low for effective views of the planets. The Moon was a special treat. Its colored fringe did not bother Tombaugh as he relished in his first telescopic look at a whole new world, with plains, craters, mountains—no longer a thing in the sky, but a place.

By the time Clyde was fourteen, in 1920, Muron and Lee shared a Sears Roebuck refractor with a $2\frac{5}{16}$-inch-diameter lens and a power of forty-five. By arrangement, the telescope would spend half of every month at one house, then move to the other. For first looks, such a telescope could open avenues that books could only hint at. For a few years, as Clyde's only telescope, it had intensive use on many clear nights. Like most Tombaugh instruments, this small telescope survives, in fine condition except for its loose drawtube. In fact, it went through another life. Long after it had outlived its usefulness as a first telescope, it was installed as a finder on his 16-inch.

Two books his uncle Lee gave him, both by an astronomer named Chambers, aided Clyde's early search for planets. One was about eclipses, and the other concerned the Sun and its eight planets. Not long afterward, Tombaugh bought a grade-school pamphlet—for a nickel—about the mysteries of Mars. Written by Latimer J. Wilson, a Nashville, Tennessee, amateur astronomer and telescope maker, the short essay was one of a series that discussed aspects of the sky. With his self-made 11-inch telescope, Wilson had produced some fine drawings of Jupiter. In the Mars pamphlet he considered the possibility that Mars could be inhabited, a conjecture that added fuel to Tombaugh's passion. Not only did Mars have a geography, but there might be creatures there to appreciate it! By the way, the Martians, according to Wilson, evolved according to the planet's lesser gravity to be twice as tall as humans. Another idea had the Martians somehow striding across the polar caps to get water for the canals. "That was a romantic idea, and I had a feeling of great awe at the knowledge of the Martians." Perhaps a tall, pole-roaming Martian might even be observing Earth one of those evenings in 1918, but Earthlings could not see the creature; with Mars in the evening sky and receding from Earth, Tombaugh's first view of another major planet showed nothing more than a reddish disk of light.

Two years later Mars was opposite the Sun again, "in opposition," and closer to Earth, shining more brightly than it had for Tombaugh's earlier look. The difference was stunning; through the telescope his father and uncle had just purchased, the planet now appeared not as an undistinguished ball of light but as a world with a polar cap and dark features spread across its surface.

Burdett, Kansas

By 1922 the Illinois farm the Tombaughs had rented was not doing well. After some years of bad weather, and a poor corn crop in 1921, the family was almost out of cash. Clyde's uncle had bought a farm in western Kansas, and after considerable thought the Tombaughs decided to take up farming there. They moved in August 1922 and spent the rest of that busy summer preparing the farm's two hundred fifty acres for wheat seeding in the fall. Years later they bought the property.

Leaving Illinois was a difficult prospect. Clyde had enjoyed the company of his double cousins and would miss them. Moving to Kansas, however, would mean new experiences, new neighbors, and a better sky to replace the clouds of Illinois. Among the routes they chose during their trip was the "Pike's Peak Ocean to Ocean Highway," marked by the standards of the time, not route numbers but roadside poles with red and white bands.[11] Muron did not transplant his threshing company to Kansas, where the threshing was done by outside contractors. In a sense this change was a blessing, for Muron had all but worn himself out running the Illinois company virtually singlehandedly.

By the following year the nation was preoccupied with the "Teapot Dome" scandal involving Interior Secretary Albert Fall of President Warren Harding's administration. The trouble began after Fall arranged for some land (Wyoming's Teapot Dome) held as reserves by the Navy Department to be transferred to the Department of the Interior. Fall then allowed two oilmen, Harry Sinclair and Edward Doheny, to develop part of the land in return for a personal loan of one hundred thousand dollars. A Senate committee finally broke the scandal, and Fall was sentenced to a year in prison.

Clyde's lively interest in Teapot Dome was a precursor to a lifelong fascination with politics. Among other worldly matters, the scandal may have occupied his mind during times he spent alone farming or dark-adapting before observing: "No matter how hard he worked during the day," Roy recalled, "if there was something to see at night, he was up. If he particularly wanted to see some critical thing, he would go into a darkened room and stay there for an hour and then go out and look through his telescope. . . . He just sat there; he couldn't read anything, and we didn't have a radio or phonograph." Clyde's mind was always so active that he had no trouble keeping it occupied during such a long wait. For instance, one day after farm work he calculated the number of cubic inches in Betelgeuse, the bright star on Orion's east flank: his solution was 1 duodecillion (39 zeros).

On the Kansas farm there was as much work as in Streator, but it was different. There may have been no threshing company, but there was threshing and there were horse-powered plows. Clyde had an almost Gulliverian love for horses, and since he spent a great deal of time with these animals, he got to know them as individuals. His respect for the horses was sorely tested one afternoon when one of the animals got the better of him. Part of a team of six pulling a plow, the horse began biting the neck of its neighbor, who jumped in annoyance. After repeated bites the other horse was wincing in pain. Tombaugh tried to get the first horse to stop, but to no avail. Getting angrier by the minute, he yanked the horse by its bridle, grabbed its neck, and threw it on the ground. "I was so mad I couldn't see straight!" Both Clyde and the horse were surprised—the horse got up and behaved very well the rest of the day, and a remorseful Tombaugh returned later to pet him.

Moving to Kansas did not mean forsaking the Sears refractor, since Lee had turned his share in it over to Clyde's family. The move also did not mean forsaking the corn harvest. The 1925 season was poor in Kansas, and to make a little money Clyde returned to Lee's Illinois farm that chilly fall for husking corn. Once again there was frost, deep mud, and cold wetness, but this time there was also a little money. After the season was over Clyde used some of these funds to buy glass, carborundum pitch, rouge, and other materials to make an 8-inch telescope.

TELESCOPE MAKING

In 1924 Tombaugh read an article in *Popular Astronomy* by Latimer Wilson called "The Drift of Jupiter's Markings."[12] It included drawings of Jupiter with features moving across its disk as it rotated. The sketches were especially moving because Wilson had created them using his self-made 11-inch reflector. Tombaugh wrote the famous amateur astronomer and he responded, offering advice on grinding telescope mirrors and referring Clyde to yet another article. Despite the brevity of Wilson's publications, they were concise, offering a lot of telescope wisdom. Wanting to do drawings like Wilson's, wanting to see the planetary geography the way Wilson saw it, Tombaugh set about making his first telescope.

In 1925 two important events occurred around the time of Clyde's graduation from high school. One was at Kansas State University, where Clyde and two other students competed with several hundred top Kansas students in exams; Tombaugh placed fourth in physics.[13] The second event he did not even know about until later. It was happening near Springfield, Vermont, at Stellafane, Shrine to the Stars, the name of both an observatory and a meeting that was to start the following summer. Alfred Ingalls, of *Scientific American*, was gathering material for the November issue's cover article about telescope making. Its title, "The Heavens Declare the Glory of God," from Psalm 19, came from an inscription on the gable of Stellafane's clubhouse. The article ignited the telescope-making movement across the country and was a powerful influence on Tombaugh's telescope making.

This was a healthy time for interest in all the sciences; in the South a religious fundamentalist movement was losing ground to a nation getting more and more involved in scientific advances. In 1925, as Tombaugh's interest in astronomy was rapidly dominating his life, the Scopes "monkey trial" was running its course in Dayton, Tennessee, where John Scopes had persisted in teaching organic evolution against state law. Defense lawyer Clarence Darrow and fundamentalist William Jennings Bryan, witness for the prosecution, made headlines for weeks. Even though Darrow lost the case in trial, he pressed the argument for evolution so eloquently that Bryan seemed the loser to an incoming tide of scientific awareness. This event must have provided interesting evening conversation at the Tombaugh dinner table and at the eyepiece.

After graduation Tombaugh set about making his telescope. Completed in 1926, it consisted of a seven-foot-long narrow wooden box, an 8-inch mirror, and a mounting with two wooden setting circles to help in pointing. He set up a grinding stand outside on a post dug into the ground south of his house. Tombaugh knew the mirror was not good. He had used a Foucault knife-edge test to judge the accuracy of the mirror's curve, but air currents in the heated house made the test inaccurate. Like almost all amateur telescope makers, he had no facilities for silvering, so he sent the mirror to Napoleon Carreau's telescope firm in Wichita. "I have ground, polished, and figured an 8-inch speculum for a reflector," Tombaugh wrote Carreau on April 26. "The speculum has a focal length of 84 inches, which I know is longer than most make for that aperture, but I wanted one of longer focal length."[14] Understanding that the long focal length was for planetary observing, Carreau responded that the mirror had a poor figure and would not work well.[15]

CELLARS AND TELESCOPES

"Everything I did was wrong," Tombaugh agreed in describing what is usually the case for anyone's first mirror. To improve testing conditions, he asked his father for help in building a "cyclone cellar" that could store dairy products and canned foods and, of course, be available as a tornado shelter. Typically, these cellars were entirely underground, except for the protruding arched roof. Tombaugh's ulterior motive for his family cellar required that it be made long enough to be an ideal testing chamber for mirrors, with quiet and cool air. After the 1926 harvest Clyde dug a pit twenty-four by eight feet and seven feet deep. Neighbors helped with the pouring of twenty cubic yards of concrete for walls, floor, and even an arched roof, stairs, and windows.[16]

With the help of the *Scientific American* article, Tombaugh began work on a 7-inch telescope for his uncle Lee. This second mirror was tested in the pristine conditions of the concrete cellar and was completed in May 1927. When Carreau silvered it, he reported that the figure was much better.[17] Getting a good response must have been a pleasure, for by now Tombaugh was taking amateur astronomy even more seriously.

Just after the new telescope was finished, a bright periodic comet returned. Visible easily in the evening sky, Comet Pons-Winnecke 1927 VII was putting on its best show since its discovery by Pons in 1819. At magnitude 3.5 and with a tail of one degree,[18] Pons-Winnecke was a memorable first comet sighting for Tombaugh. Its brightness and size made it an ideal comet for a telescope, for its coma and most of its tail were fully visible in one or two fields of view.

THE 9-INCH

A month later the telescope was shipped off to Uncle Lee, and with the money he earned for it, Clyde set about his third project, a 9-inch. In August 1927 he purchased the glass blanks, and in September he set up with a stand on top of which rested a blank piece of glass called the *tool*. Tombaugh worked the mirror blank on top of the tool in a process that involved several steps, the first being rough grinding with carborundum to give the future mirror a concave surface. Next came fine grinding with finer grades of carborundum. Then he prepared a lap of pitch, a derivative of pine wood, to attach to the glass tool for the polishing process, which is designed to refine the accuracy of the mirror's curve. Using polishing rouge (actually, heated ferric oxide), he polished the mirror.

Finally came figuring, in which the polished mirror surface was repeatedly tested and altered until a perfect paraboloid surface was reached, free of errors such as zones that result from a lack of randomness of one's stroke. "I struggled with the figuring of that mirror for many weeks, and by trying to get one zone out I'd generate another. Finally, I got them all licked and obtained an excellent parabolic curve on the mirror. That really paid off, for it would take magnifying powers as high as four hundred." The telescope's mounting included parts from Muron's retired 1910 Buick and a cream separator.[19]

In the winter and spring of 1928 Tombaugh progressed on his 9-inch telescope. Impressed by Tombaugh's work, Napoleon Carreau wrote that his company would need an assistant for telescope making, but not for a year or so. Meanwhile, this was the time to observe and to show off the sky to neighborhood friends. By the fall of 1928 the excellent telescope was completed, and at night Tombaugh drew what he saw on the disks of Mars and Jupiter. "I remember staying up most of a night watching Jupiter, and within an hour I was aware that the markings were drifting across the disk! That was a big thrill for me, for I was watching a planet rotate on its axis."

Turning the declination circle some degrees north and adjusting the hour angle, Tombaugh found Messier 13 in Hercules entering his field of view. Consisting of many thousands of distant stars, it is one of the most majestic objects in the sky. No. 13 in Charles Messier's catalog of "nebulae" is the best of the northern hemisphere's globular clusters, and for Tombaugh it was not the diffuse blob of light that the old refractor would have shown, but a central glow surrounded by uncountable stars.

Globular clusters were hot subjects in astronomy at that time, for Harlow Shapley, of Harvard Observatory, had earlier announced the results of his research on their distances. Since they obviously consist of many stars, they appear to be distant members of our galaxy and thus could define the galaxy's dimensions if their distances could be found. These clusters contain stars known as Cepheid variables, after Delta Cephei, that vary in brightness over a fixed cycle. Shapley and Henrietta Leavitt had determined that the periods of these stars were directly proportional to their intrinsic brightnesses. By finding some Cepheid variables in these clusters and comparing their apparent magnitudes, influenced by their distance from Earth, with their actual brightnesses, Shapley could determine their distances.

One of the greatest discoveries ever made in astronomy, this "period-luminosity relationship" allowed astronomers to conclude that the globular clusters appeared to range from twenty-two thousand to more than two hundred thousand light-years away. This incredible find indicated that the Universe was far larger than anyone had believed. Shapley's work lit the popular imagination and was reported in magazines such as the *Literary Digest*, which Tombaugh read. To be able to walk out the front door and see one of these clusters through a telescope was a deeply thrilling event.

In short sections in most issues the *Literary Digest* summarized the astronomical picture that was changing so radically in the 1920s. "I will never forget how revealing those articles were," Tombaugh remembers. "Here was a Universe vastly greater than anyone thought. Can you imagine what an awful shock it is to revise your thinking about the Universe to enlarge it so much? And I saw it happen in my lifetime."

The Hailstorm

Although the Tombaugh farm concentrated on wheat, in 1928 Muron planted about fifteen acres of oats. The fine spring weather seemed to promise a good crop, and one evening Muron told his son Roy, "If that doesn't make fifty bushels to the acre, it won't make a one." This was good news, for a healthy farm meant that Clyde might be able to afford to start college in the fall.

June 20, 1928, was hot and muggy, with a stiff breeze that provided little relief from the oppressive heat. By midafternoon, ominous storm clouds had gathered in the northwest. The wind picked up dramatically, and the family rushed inside before torrents of rain fell amid thunder and wind. Then the hail came. It lasted only fifteen minutes but destroyed the wheat crop and the promising oats as well. Muron had been right: the oat crop didn't make a one.

"Dad had some philosophies about dry land farming that he shared in later years," Esther recalled. "He could never, he said, criticize persons for gambling; a dry land wheat farmer was the biggest gambler there is. Another time he said, 'My occupation is a wheat farmer, but I milk cows and raise chickens for a living.'"[20]

Looking across the land, Clyde reflected on the vagaries of weather, which had spared land just a few miles away. College was now out of the question for the immediate future. Farming, Tombaugh mused, was for fatalists; he wanted a life where he would have more control. For the rest of the summer of 1928, Clyde ran his neighbor's new combine on a field just outside the hail strip. This was the year Kansas farmers began to switch to these more efficient machines, which would harvest and thresh, saving an enormous amount of work. Because of the storm the Tombaughs did not buy their own combine until 1929.

One day, when working on Elmer Steffen's farm, he noticed that the ten-thousand-bushel grainery was beginning to fill up. Using long-hand arithmetic, he calculated the number of kernels of wheat in the entire grainery. "Clyde," Steffen asked, "why are you working in wheat fields when you do things like this?" This thought was not far from Tombaugh's mind, and now it was time to do what he wanted to do, to follow the advice of his neighbor and get into science. "I wanted to get into some kind of a job where making money did not depend so much on the weather," Tombaugh said of his desire to be a professional observer.

Late in December, Tombaugh mailed his best drawings to Lowell Observatory, in the hope that the astronomers there would give him some suggestions. An answer arrived quickly, but from an unexpected direction: instead of giving him advice about his drawings, they asked about his training, and in a further letter they inquired about his physical health and ability to handle long cold nights in an observatory dome. A position was open, it seemed, for someone to operate a new photographic telescope to take long-exposure photographs of the sky.

The observatory's need for a dedicated amateur astronomer who would be willing to work for long hours at low wages exactly coincided with Tombaugh's urgent desire to get away from farming. In a letter to the observatory trustee, Roger Lowell Putnam, V. M. Slipher showed his concern about hiring a professional astronomer with his own agenda of things to do; it seemed to him that this "young man from Kansas"[21] was good enough to do the work without causing too much of a drain on the observatory's resources.

It was time to decide between Lowell's offer and Carreau's offer of a position in telescope making. It is a fact that amateur astronomers naturally divide into those who make telescopes and those who use them most effectively, since the various skills are different. Tombaugh was proficient in both areas and had the rare good fortune of having offers for paid positions in both. In his heart there was really no choice. What young amateur astronomer would pass up a chance to use a professional telescope? The mysterious new program at Lowell Observatory was far superior to a job at a telescope-making firm. The decision had enormous consequences. "I went west instead of east. It makes me shudder how my life hung in the balance at that time. If I had gone to Wichita, I would have been an obscure telescope maker and no one would ever have heard of me."

Had that decision gone the other way, no one would have heard of Pluto either, possibly for another fifty-six years. In May 1986 a competent observer accidently photographed the planet on two nights and recognized its unusual motion without immediately identifying it as Pluto. Undoubtedly, he would have followed it up. "The curious thing," notes Brian Marsden of the Central Bureau for Astronomical Telegrams, "was that Clyde had written that very month in *Mercury* magazine that 'had it not been for Percival Lowell, I am confident that Pluto would not have been discovered, even unto this day!' " [22] Marsden quotes another "discovery" of Pluto, reported by Liisi Oterma in *Turko Informo* No. 6: "The planets are readily recognized even with rather short intervals between exposures. For example, we once discovered Pluto on one of our double point plates that had been taken with an interval of 53 minutes." [23]

What would become of his Kansas home in the future? Although the farm remained in the Tombaugh family, years later Clyde sold out his interest to his brother Charles. He did help pay the mortgage and taxes during the depression years, for at his observatory salary of $125 per month Clyde was the family's "rich aristocrat." That was all to come; now everything was uncertain.

Departure day, January 14, 1929, was difficult. Clyde had to invest in his own railway ticket, funded with money earned from running the neighbor's combine, and since the Pullman fare was expensive, he faced the twenty-eight-hour journey in a coach car. His mother gave birth to Rachel (now called Anita Rachel) just twelve hours later while Clyde was on the train, and he did not see his baby sister for six more months. "I was upset about that. I knew that my mother was about to have the child, but I just needed to get out there. I didn't want to spoil my chances with Slipher." The hugs were over, and Clyde was on board. He wondered about the fact that the position was for only a ninety-day trial period—what if it didn't work? He didn't even have the funds for a ticket home, although one would hardly expect Lowell Observatory to leave him abandoned on some railway track because of two or three trailed exposures.

With a sharp jolt the train inched from the station. Slowly at first, then picking up speed, it carried Tombaugh into a new phase of his life, in which memories of the rhythmic clanking noises of the thresher and its steam engine would damp out to the soft click of a blink comparator.

Chapter 3

URANUS, NEPTUNE,
AND PLANET X

Uranus

IN a historical sense, planet finding began with William Herschel, and Herschel had begun his adult life not as an astronomer but as a musician. Chronically out of funds, he once traveled to Genoa from Yorkshire but had absolutely no money to get back to England. To solve the problem, he gave a concert by himself, playing a harp and two horns at the same time. Try to imagine the future discoverer of Uranus, one of the greatest astronomers of all time, with one of the horns fastened to his shoulder so he didn't have to hold it! Even if we cannot picture such a sight, the people of Genoa certainly did, for they came in droves and paid his way back to England.

At this period of his life William Herschel's highest ambition was to establish a reputation as a musician and composer. In order to get from place to place he had to buy himself a horse, and he frequently rode more than fifty miles a day across the moors in all kinds of weather. When he got back to his house in the evening, he was so exhausted that he could do no further work and had to spend the evening in relaxation, which for him was studying Italian, Greek, and Latin, as well as music and mathematics. "The theory of music being connected with mathematics," Hershel wrote, "had induced me very early to read in Germany all what had been written upon the subject of harmony; and when, not long after my arrival in England, the valuable book of Dr. Smith's Harmonics came into my hands, I perceived my ignorance and had recourse to other authors for information, by which I was drawn from one branch of mathematics to another."[1]

During all these years of music, Herschel's hours, one might think, were so full and so fulfilled that he could have had no further ambitions. But that was not the case, as we note in this innocuous journal entry for May 10, 1773: "Bought a book of astronomy and one of astronomical tables."[2] Something, possibly deriving from his interest in music theory through mathematics, had ignited an interest in astronomy. In May he procured some short object glasses and had tubes made for them, beginning with one four feet long:

> With this I began to look at the planets and the stars. It magnified 40 times. In the next place I attempted a 12 feet one and contrived a stand for it. I saw Jupiter and its satellites with it. After this I made a 15 feet and also a 30 feet refractor and observed with them. The great trouble occasioned by such long tubes, which I found it almost impossible to manage, induced me to turn my thoughts to reflectors, and about the 8th September I hired a two feet Gregorian one.[3]

Soon Herschel was making his own reflectors, using speculum metal for reflecting surfaces made from copper, tin, and antimony. Mirrors made of that material survive more than two hundred years later, their fragile coats still capable of a skyward look. Around this time he also moved into a larger house in Bath, bringing his sister Caroline, who had begun to assist enthusiastically in his work.

The observing yard at No. 19 New King Street, from which William Herschel made some of his important early discoveries, was small. His program of observation was essentially a patrol, or exploring, operation during which he recorded carefully everything he could see. Late in the evening of Thursday, March 13, 1781, only eight years after he had begun his love of the stars, he observed the region of the star h Geminorum: "I perceived one," he wrote later, "that appeared visibly larger than the rest; being struck with its uncommon magnitude, I compared it to h Geminorum and the small star in the quartile between Auriga and Gemini, and finding it so much larger than either of them, suspected it to be a comet."[4]

Never before had such an object been found. Although it showed a perceptible disk, it resembled more a planet than a comet, and intensive observations over the following five months revealed that its slow motion indicated great distance. Finally, on August 31, the mathematician Anders J. Lexell published an orbit that showed the object's closest distance from the Sun as sixteen astronomical units, sixteen times the distance between Earth and Sun. Well beyond the orbit of Saturn, this was a new planet, the first discovered in recorded memory.

The discovery caused a sensation. At the end of 1781 Herschel was offered the Copley Medal of Britain's Royal Society. Apparently, after some persuasion, he did attend the ceremony, although he felt that he would much rather be building telescopes than accepting medals. The citation was astonishing in its accurate prediction of future discovery: "Your attention to the improvement of telescopes has already amply repaid the labour which you have bestowed upon them; but the treasures of the heavens are well known to be inexhaustible. Who can say but your new star, which exceeds Saturn in its distance from the sun, may exceed him as much in magnificence of attendance? Who can say what new rings, new satellites, or what other nameless and numberless phenomena remain behind, waiting to reward future industry?" [5]

Five large satellites were eventually discovered, and on March 10, 1977, astronomers detected a set of rings surrounding Uranus. With Voyager 2's encounter in 1986, ten small satellites were found. Herschel went on to enjoy a highly successful career in astronomy. Discoveries of double stars and nonstellar objects called nebulae mounted with each new and mightier telescope he designed and built. With two grants of funding from King George III, he completed a huge forty-foot-long telescope. From its first light, it was a superb instrument: on its very first night of use Herschel discovered Mimas and Enceladus, two close moons of Saturn. But although he tried to use it for some time, the speculum mirror tarnished very easily. The telescope was unwieldy and spent more time in the shop than in the observing field. Extremely difficult to enjoy, its use required the observer to stand on a little platform high above the ground. It began to fall into disuse, and all that now remains is the bottom of the tube.

Neptune

On May 8, 1795, only fourteen years after Herschel found Uranus, a prominent French astronomer named Joseph Lalande was mapping fields of stars. Two nights later, while checking his plots against the stars in his telescope's field of view, he caught what he thought was an error he had made in mapping. Correcting his chart, he moved on. Apparently, Lalande never revisited that field, never noticed that his misplotted star would continue to shift along the background of stars.

The idea that a more distant planet was disturbing the orbital motion of Uranus gathered momentum as soon as astronomers began to collect prediscovery observations of Uranus—older plottings by astronomers unaware of what it was. In 1834 the amateur astronomer Rev. T. J. Hussey wrote to astronomer George Airy suggesting that some distant body might be affecting the orbit of Uranus. Airy's response to this intriguing idea was the first of a remarkable series of unfortunate decisions that continued for twelve years. Conceding that the gravitational influence of another body might indeed be at work, Airy believed that such a body would be too hard to find, and he promptly forgot about it, as apparently did everyone else, for seven years.

The next figure in the Neptune saga was John Couch Adams, a twenty-three-year-old student at Cambridge. The problem of the orbit, he thought, could be solved by the presence of an undiscovered body, and after his graduation he sent his calculated position for the new object to John Challis, his professor of astronomy and a scientist well known for his comet observations. Challis in turn sent the work to Airy, who by this time was England's Astronomer Royal and director of the Greenwich Observatory. Again, Airy expressed considerable interest but did not encourage a search. Even though Airy has been criticized for this response, one wonders why no one else mounted a search at the time. After all, several large telescopes throughout England were now fully capable of looking in the area of Adams's solution to see if there indeed was a planet there. Other than Airy and Challis, not many people even knew about these predictions; perhaps Adams should have published them.

A man of great genius and intense discipline, Airy was a difficult person to deal with. Under his regime an observing assistant underwent a torturous and unbreakable three-day routine, involving, on the first day, twenty-one hours on an instrument called a transit circle. Day two was light, consisting of a small amount of computing, but day three followed with all-day computing and, depending on weather and the phase of the Moon, all-night observing. Day one of a new cycle would begin immediately after the night's observing, at six in the morning![6]

It is possible that Airy's preoccupation with such an unbending routine prevented him from committing his observatory's telescopes to an early search for Adams's predicted planet. It is more likely, however, that the busy Astronomer Royal was not sufficiently persuaded by Adams's evidence. Today someone in Airy's position might receive several discovery claims each week, and he or she must decide if any such claims are worth pursuing. It is really up to the discoverer to provide enough evidence to encourage a director to investigate a claim. Nevertheless, Airy did eventually show an interest in Adams's work and wrote Adams with some questions. Angered by what he perceived to be a lack of understanding on Airy's part, Adams never answered the query. In the meantime, Challis was also aware of the prediction but did not bother to look for the possible planet.

Meanwhile, in France, Urbain Jean Joseph Leverrier was about to complete his own calculations. Unaware of Adams's work, Leverrier by the end of 1845 had calculated the difference between predicted and actual positions of Uranus and published a paper that impressed Airy. Months later, Leverrier sent Airy actual predictions for a new planet; Airy noted the agreement to within one degree, twice the Moon's diameter, of Adams's solution. In his response to Leverrier, Airy noted this remarkable agreement but inexplicably never told Adams about it. Now Airy encouraged Challis to start a search, which he did, but instead of going right to the predicted position he worked inefficiently, if thoroughly, across a ten-by-thirty-degree strip of sky.

Leverrier also had difficulty mounting a search at the Paris Observatory, so he asked Johann Galle at Germany's Berlin Observatory for help. Accepting Leverrier's proposal, observatory director Johann Encke permitted an immediate scan. On the night of September 23, 1846, Galle and a young student named Heinrich d'Arrest opened the observatory dome. Going directly to the predicted position, they found nothing, but then Galle called out the positions of star after star while d'Arrest checked a chart. All was as it should have been, except for an eighth-magnitude object d'Arrest could not identify on the chart. With mounting excitement the two men looked at the object, then at each other, then summoned the director. Encke carefully measured a small disk 2.2 arcseconds across. A new planet had been discovered.[7]

Four years after the discovery of Neptune, and its moon Triton in the same year by William Lassell, the English poet Alfred Lord Tennyson included this stanza in his epic poem "In Memoriam":

Is this an hour
For private sorrow's barren song,
When more and more the people throng
The chairs and thrones of civil power?

A time to sicken and to swoon,
When Science reaches forth her arms
To feel from world to world, and charms
Her secret from the latest moon?

Back in England, Challis's cumbersome search had actually gone over the planet twice without noticing it. Only four days after beginning his search, the telescope came across Neptune, but Challis did not reduce these observations until after the discovery. Once the discovery was announced, Airy suggested that Adams should also be credited with the discovery, a move that infuriated the French and launched a huge controversy. By June 1847 the affair had died down, Adams and Leverrier finally met, and they became lifelong friends.

In retrospect, both Airy and Challis were overly criticized for the Neptune scandal. Both were top scientists who paid dearly for their failure to take Adams's calculations seriously until it was too late. More important, their loss was more political than scientific, since the planet was found quickly after the final calculations were published.

Planet X

Half a century later a dreamer named Percival Lowell started building a private observatory in Flagstaff, Arizona, to study Mars during its favorable viewing season in 1894. Lowell began with a borrowed 18-inch diameter refractor that enabled him to observe long "canals" — actually long and narrow dark streaks — across the planet, in a network that included dark oases where the canals intersected. In 1896 the 18-inch refractor was replaced by a 24-inch. Because Lowell's observations were not confirmed by many other observers, they were ridiculed during his life, as were the conclusions he drew from his work. Mars, he thought, was inhabited by a mature population with engineering expertise far ahead of ours. A pacifist in politics, Lowell wrote that the canal complex could have been completed on this planet-wide scale only by a civilization that had united to solve its common need for water.

Although Lowell never gave up his dream of an inhabited red planet, he was bitterly disappointed with the controversy his canals had caused. He next sought something new. Since the orbital motion of Uranus was not totally explained by the presence of Neptune, could there be an elusive Planet X? It seemed reasonable at the time to suppose that another planet might be still undiscovered. Setting out much the same way Adams and Leverrier had, Lowell calculated the position based on orbit discrepancies, and by 1905 his first search had begun. Using a 5-inch refractor telescope made by John Brashear, one of the finest optical craftsmen of his time, E. C. Slipher took a series of photographs at five-degree intervals along the mean plane of the orbits of all the known planets, the *invariable plane*. Although this small telescope had a substantial field of view of about five degrees (about ten Moon diameters), it could reveal stars barely to the sixteenth magnitude. Even if it could have gone deeper, it would have missed Pluto, which was too far south of the invariable plane at the time.

This grueling search involved taking and searching through four hundred forty photographic plates, exposed for about three hours each. Lowell would examine these plates for any moving object by laying one plate over another so that all the images were offset slightly. With a hand magnifier he would then look for any object that was not starlike or that moved. One set of these plates caused excitement and some embarrassment to the observatory when Lowell reported a comet discovered by Slipher. Intense searches from observatories at Bamberg and Heidelberg did not confirm the comet. Unfortunately not learning from this error, the observatory later reported a second comet with two tails, found nearby on the same set of plates. The observatory's failure to confirm these objects through additional photography is a sign that this first search was not well executed.[8]

After two disappointing years this search stopped. In 1909 Lowell mounted a second search, this time with a new 40-inch reflector telescope. The large mirror of this telescope meant that instead of three-hour exposures, seven-minute exposures could go faint enough. Unfortunately, however, the field of the telescope was so small that a great many plates would have to be exposed to cover large amounts of sky. Each plate covered a field of about one degree. Examining the new plates was easier now, too; by 1911 the observatory had acquired a Carl Zeiss blink-microscope comparator, a mechanical contrivance set up so that two plates could be examined at the same time, first a small area on one, then the same small area on the other, with a motor rapidly switching the optical path. Instead of using a hand magnifier, the observer could look at the plates in greater detail and in greater comfort through the eyepiece of this marvelous instrument.

The search program did not match the abilities of the new instrument. Looking for the planet meant searches in Gemini far from the opposition point, so that a slow-moving asteroid could easily be taken for the missing major planet. Instead, Lowell relied heavily on mathematical calculations pointing to a particular area, and when nothing promising was found, he would move to another area.

This search lasted until 1912, when Lowell borrowed a 9-inch wide-angle camera from Swarthmore College's Sproul Observatory, an institution still famous for its many fine wide-angle photographs from which star charts have been prepared. By 1914 the camera was set up, and upward of one thousand plates were taken and examined on the blink comparator. Carl Lampland was in charge of this third search, and two observers, T. B. Gill and E. A. Edwards, took most of the plates.[9]

This program actually recorded Pluto in 1915! At the edge of two plates, taken on March 19 and April 7, the frozen dark speck of Lowell's dreams remained, but no one noticed. At the time these images were taken, the region was far from opposition. In this rich area of the Milky Way the plates were so crowded with stars that the moving planet was missed. Even if it had been found, its motion could hardly have been distinguished from that of an asteroid near its stationary point. Even though Lowell had recommended that the plates be taken at opposition, that procedure was not being followed during much of this third search.[10]

With the Mars observations in disrepute, the search for Planet X going nowhere, and World War I challenging the philosophy he held dear, Lowell became thoroughly depressed. On November 12, 1916, he had a stroke and died. The search stopped, and the telescope was returned to Swarthmore College.

The next fourteen years were troublesome for the observatory. Although Lowell had left his institution well funded, his widow, Constance, sued to break her husband's will, a strategy that kept most programs on hold at a time when much of the rest of the country's economy was flourishing. Handling the observatory's interests as trustee until 1927 was Guy Lowell, who was succeeded by Lowell's nephew, Roger Lowell Putnam. During this time the scientific work continued, especially V. M. Slipher's seminal investigation of the nebulae (see Chapter 7). After the suit was settled a decade later, Putnam strongly encouraged the two Sliphers and Lampland to resume the planet search. A new 13-inch camera was funded by A. Lawrence Lowell, president of Harvard University, and built by Carl Lundin, of the Alvan Clark firm. Late in 1928 Clyde William Tombaugh, whom Slipher described in a letter to Putnam as a young man from Kansas with enthusiasm and the ability to spend long hours in a cold dome completing good observations, was hired to conduct the fourth search for Planet X.

Chapter 4

THE FIRST YEAR

A RE YOU Mr. Tombaugh?"
With these words Vesto Melvin Slipher met Clyde Tombaugh at the Flagstaff depot on January 15 after his long train ride. Feelings of excitement and anxiety had taken hold as the train approached Flagstaff. A thousand miles from home, Tombaugh was about to start work at a famous observatory. He was there at his own expense and on the advice of letters from a man he had never met.

The Lowell Family

Had Uranus and Neptune not been on the same side of the Sun around 1822, causing the pull of one to affect the observed orbital motion of the other, no one would have thought to search for another planet, Neptune's discovery would have been indefinitely delayed, and Percival Lowell would probably never have thought of searching farther out.

Lowell was not the only scientist actively engaged in a planet search, although his methods were unique in that they were both theoretical and observational. In addition, William Pickering and Thomas Jefferson Jackson See had both made predictions of distant planets. Pickering noted several, including "O," whose orbit was close to that of Lowell's "X" but whose magnitude was fainter, and "P," whose orbital elements changed radically from one prediction to another and whose final predicted position was somewhere in Aquarius. Although none of these

39

planets was ever found, Pickering did claim that Pluto was his Planet O. See's planet, already named Cronus, also was never found.[1]

Percival Lowell came from one of the country's most distinguished families, even mentioned in John Bossidy's famous toast, "On the Aristocracy of Harvard":

> And this is good old Boston,
> The home of the bean and the cod,
> Where the Lowells talk to the Cabots,
> And the Cabots talk only to God.[2]

The Lowell family was rich, intelligent, influential, and occasionally eccentric. The generation before Percival's included James Russell Lowell, born in 1819, who wrote poetry, essays, and satire. Percival was born March 13, 1855, the day in 1930 that would ultimately be chosen to announce the successful result of his cherished planet search. His younger brother was A. Lawrence Lowell, who became president of Harvard University. His sister, Amy Lowell, was a well-known poet whose major works were published between 1916 and 1920. In the following generation Robert Lowell published his first volume of poems in 1944.

The fact that in 1928 the observatory still belonged to Percival Lowell was never lost on anyone there, and his devoted followers were continuing the legacies he left. Although Tombaugh was only ten years old when Lowell died, the founder's presence was so strongly felt at the observatory that shortly Tombaugh felt as though he had known him. One of these legacies was his widow Constance, who visited the observatory each summer for about a week. In addition to delaying the observatory's programs by contesting Percival Lowell's last testament, Mrs. Lowell objected to the director appointed by her husband's will, V. M. Slipher, a man she detested. (Her choice was Samuel Boothroyd, of Cornell, who had assisted Lowell years earlier.) Tombaugh suggests that Slipher's continuing work on the red shifts of the distant nebulae may have angered her as well; she felt that her departed husband had already "found out everything worth knowing. She didn't like the idea of new things being done that weren't done by Lowell." Eventually, the case was settled out of court, but the legal expense to the observatory was enormous. Roger Lowell Putnam was now trustee, and by April 1927 the remaining issues had been settled.[3]

First Days

Quite a bit of melting snow was on the ground that first day, and the road up to the observatory was icy enough that V. M. Slipher did not make it to the top on the first try; he had to back down to try again.

The observatory staff was smaller than Tombaugh had expected. Dr. Slipher was slow to initiate a social conversation, although he would always be available to talk with Tombaugh when needed. A pioneer of planetary radiometry, Carl Lampland worked in this field at the same time S. Nicholson and E. Pettit were doing similar research at Mt. Wilson. During his conversations with Lampland, Tombaugh would become acutely aware of the scientist's near-reverence for Lowell. In fact, of all the people at the observatory, he was the most courteous to Lowell's widow during her tense summer visits. He and Tombaugh struck up a friendship as soon as Tombaugh arrived.

The first full day at Flagstaff began with a ride to town with Mr. Jennings, the handyman. He left Tombaugh off at a café and then went to pick up the observatory mail. He returned around the same time Tombaugh finished his breakfast, and the two drove back up Mars Hill, this time with a little more speed than Slipher had used the afternoon before. At midmorning Slipher asked Jennings to show Tombaugh the new 13-inch astrograph. The two men walked up the snowy path to the dome, the first of thousands of walks Tombaugh would take between the two buildings.[4]

Inside the observatory that January morning, instrument maker Stanley Sykes and his son, Guy, were still installing the hand controls, as a finishing touch on the instrument. The welds were still rough, and the scope was unpainted; finishing that was Tombaugh's job later.[5] Sykes's career at the observatory eventually spanned half a century. Tombaugh compliments his work: "It was a darned good instrument. It was made very heavy so that gusts of wind wouldn't shake it. It was easy to operate so that those long exposures wouldn't kill you. I feel we owe a great deal of our success to this well-built telescope."

Visitors to Lowell Observatory have taken the public tour, offered weekdays, almost since the observatory's founding. That afternoon Clyde Tombaugh attended the tour, in this case given by Mrs. Fox, the secretary, and a few days later he began conducting the half-hour tour that began at 1:30 in the library and continued up the path to the 24-inch refractor. It was to be his regular duty for many years, including the crowded days after Pluto's discovery. In May 1967, as a teenager getting serious about astronomy, the author took the same tour. The guide did not identify himself until someone asked his name: "Robert Burnham." "You mean the fellow who found six comets?" We were most impressed that such a busy and well-known staff astronomer took time to talk to the public.

During the first week Tombaugh was also assigned the furnace detail. The dirty chore didn't bother him a bit. V. M. Slipher showed him where the split pine logs were stored and instructed him to throw several into the furnace every two hours. To keep the fire going through the night, the last stoking would include some one hundred fifty pounds of coal that would sit between the logs and the furnace door, keeping the furnace warm enough for easy rekindling the next morning when handyman Jennings arrived. On observing nights the Sliphers or Lampland took over the furnace task.[6] Another weather-related concern was that Flagstaff's heavy snows would affect the delicate dome structure of the 42-inch reflector. Used mostly by Lampland, the reflector had an odd dome of wood frame covered by chicken wire and canvas. Being careful to step only on the supporting beams, Lampland would push the snow off with a long hoe. For many years he cleared the snow himself, fearing that someone else would tear through the canvas by mistake. After the first snowfall, Lampland decided to ask Tombaugh to take care of this task.[7]

In some nonscientific matters Lowell Observatory was a democracy. Everyone except the secretary had a part in menial jobs like restoking the furnace. Conversely, when visiting astronomers came by on their way to or from Mt. Wilson, everyone including Tombaugh would have time to spend with them. Occasionally, these people would have dinner at a downtown hotel; Tombaugh was always included. With Earl C. Slipher, formerly mayor of Flagstaff and now a state senator, democracy had more literal meaning. Since the state legislature met during the first few months of each year, the younger of the Slipher brothers was not in Flagstaff when Tombaugh arrived, and Tombaugh did not meet him for several weeks. Slipher was out of town again when Pluto was discovered the following winter. E. C. Slipher and Tombaugh got along well from the moment they met; "I guess he found in me a sort of companion because I was interested in things most people wouldn't listen to him about. We had a wonderful friendship in regard to observing the planets."

The Astrograph

With the chores of tours and furnace stoking, Tombaugh was passing time until the 13-inch was ready and the photographic patrol could begin. During these days he had learned the ropes of the observatory, and the tasks he was performing, though menial, were important and drew him effectively into its world.

The 13-inch triplet lens that would become the heart of the final search for Planet X at Lowell Observatory has an interesting history. At the time of his death in 1925, Reverend Joel Metcalf, one of the finest optical workers in the United States and also a comet discoverer, had been working on this lens, and Lowell Observatory trustee Guy Lowell was able to buy it from his estate for the observatory.[8] After a long search for someone to complete the lens, the new trustee, Roger Lowell Putnam, arranged for Carl Lundin, of the famous optical firm of Alvan Clark and Sons, to fashion the optics for the new patrol telescope.

The lens arrived on February 11, only twenty-seven days after Tombaugh. The timing could not have been better; thanks to Sykes's careful work, the telescope was ready to use before the objective arrived. Everyone was uneasy at the newly completed lens by Carl Lundin, in part because trustee Putnam had already paid him more than the price originally agreed on and the lens had not even been inspected! The reason for the overpayment was that one of the lens's two elements was very thin, making the figuring process much more difficult. In an exchange of letters with Putnam, Lundin explained the difficulty and proposed a price that would still allow a profit for Clark and Sons. Putnam offered to split the difference, and Lundin immediately accepted. The check was written without consultation with the senior staff at Lowell.

On that morning, Tombaugh recalls, V. M. Slipher "was still really upset about it; he said Putnam should not have done that until we had run checks on its quality." The lens was carefully loaded into the observatory Model T truck. When the truck arrived, Lampland and Tombaugh joined the others at the 13-inch telescope building. Everyone breathed more easily when the screws were undone, the box opened, and the lens found to be in one piece. Then V. M. noticed that the angle on the focusing slot was steeper than he would have liked; a small turn would thus change the focus dramatically. Carefully, the objective, mounted in its cell, was attached to the tube. On the first clear night a focus plate was taken and the correct reading recorded. (For a focus plate, a region of stars is photographed, then the telescope is moved slightly in right ascension, the focus is changed, and the same plate is reexposed. The process is repeated through a range of focal positions.)

A few nights later V. M. Slipher, Lampland, and Tombaugh took a half-hour exposure of Orion's belt and sword. Even though the constellation was west of the meridian and in the southern sky at the time of the exposure, Slipher was familiar with this region and wanted to compare the new exposure to others he had seen.

After the Orion plate was developed, Slipher forgot his monetary worries. The images were sharp all the way across the plate. The astronomers would be able to use the big 14-by-17-inch plates, giving larger fields than standard 8-by-10-inch plates. This lens surpassed Slipher's hopes and would be more than adequate for the planet search. The only difficulty was a very slight asymmetry in star images; on bright stars it would be drowned out completely, and it was not powerful enough to notice with the faintest stars. Stars at a particular magnitude in between, however, would show enough asymmetry that astrometry—the calculating of positions of objects on the plate—would be difficult and uncertain. It was a problem noticed only when the images were examined under high power, as Lampland later complained when he was measuring images of Pluto. It did not affect Tombaugh's blinking operation.

The success of this first plate meant that the new search for Planet X could begin. V. M. Slipher designed a plate holder to handle the large plates, bending them to a "Petzval curvature," a slight plate concavity toward the lens that defines the useful field of sharply focused stars.

Around that field the star images become elongated. The plates would be loaded emulsion-side down to rest on strips of brass in the middle of the sides and ends and on gusset faces at each corner. The observer would then lock five thumbscrews on the back to force the plate to the correct curvature.[9] Tombaugh's training on the telescope went smoothly, although once he did take two hour-long exposures on the same plate. The area boundaries were penciled on a copy of *Norton's Star Atlas*, a famous work intended for amateur use and still in print in 1991.[10]

Almost two months had passed since the lens had arrived, and still there were problems in setting things up for a regular observing program. Like all equatorially mounted telescopes, this one had a mount with two axes, one in right ascension that would move along with the guide star as it crossed the sky, and the other in declination. One problem was a tendency (common to many telescopes) of the right ascension axle to track erratically at certain hour angles. By moving the counterweight slightly to increase resistance on the worm gear, Tombaugh solved the tracking problem.

Now everything seemed ready. Since July 2, 1916, when the last forlorn exposure had been taken in hopes of finding Planet X, thirteen years had passed. That last log entry read "Lunch," implying not an ending but a break.[11] Now it was the evening of April 6, 1929. Guiding on Delta Cancri, Tombaugh took a one-hour exposure. It was repeated a week later, and the fourth Lowell Observatory planet search had started.

Some interesting additional problems cropped up during this first moondark period. One involved the mount; during an exposure Tombaugh heard a chugging sound, and then suddenly the guide star was off center. Quickly, he moved it back, but the resulting exposure had a double image, making the plate unusable. When the effect happened again, Tombaugh realized that the telescope was at exactly the same hour angle it had been at before, forty-two minutes of time west of the meridian. (The meridian is the imaginary circle against the sky that passes through the zenith from north to south.) The entire telescope lurched to the west, shifting images by half a millimeter whenever it passed this point. Tightening the axle did not help, unless it was tightened so greatly that the scope would not move at all. There seemed no other solution but to cause the lurch to take place on purpose, but before the exposure began. By moving the telescope through forty-two minutes west, he would hear the chug; then gently he would bring it back to the correct position and begin the exposure. Another problem not corrected immediately was the difference in brightness of star images from one plate to another. Haze, light cloud, or poor seeing caused this difference, and only after Tombaugh got experience blinking the plates himself did he figure out how much longer to expose under different sky conditions to equalize the images.[12]

In the hope that Planet X would yield to the new telescope's power, in the spring of 1929 V. M. Slipher asked Tombaugh to photograph the ecliptic in Gemini and then go into Cancer and Leo. Slipher may have realized that even if Planet X were in Gemini, it would be so far west of opposition in April that its motion would be barely detectable, but he never told Tombaugh. Slipher was hoping that "X" would be bright enough to be confirmed somehow without using its motion as an indicator of its distance. On April 11 and 30 two plates were taken of the region around Delta Geminorum. The first of these broke into two pieces,[13] although it was still usable and could be examined. It was the first of a number of plate casualties involving this field: "I kept making Delta Gem plates because I had casualty after casualty," Tombaugh reminisces. "I bet that in '29 I made half a dozen plates and none of them were much good. I ought to have had the presence of mind to know that this was the place to find the little elusive rascal."

Except for his failure to understand the need to take plates at opposition, and to take three plates of every field (two to blink and one for confirmation), V. M. Slipher had planned this fourth search for Planet X very carefully. The one-hour exposures would yield stars down to magnitude seventeen, a full four magnitudes fainter than Lowell's predicted magnitude of twelve to thirteen.[14] When the fourth and final search for Lowell's Planet X began, the Lowell staff felt if the planet were anywhere near the predicted position in Gemini, it should be easily located with the large new astrograph. In retrospect, that was an obvious error in judgment, but a common and understandable one. The history of astronomical thought is laced with problems whose solutions are envisaged as the end result of the construction of a larger telescope. In some cases the new telescopes do solve the problem; in others the question put to the large scope is sent back with ever larger questions.

This first season provided quite an initiation. During one predawn session in May, Tombaugh was completing the last plate exposure, with the dome's open slit right over a large rock.

> Suddenly I heard this bloodcurdling growling right outside the dome. It sounded vicious, and I thought it would climb up the wall and come in through the slit! I kept watching to see if that thing would come over the seal of the door. I was petrified! If it had, I would have bolted out and started to get the heck out of there fast. The sound died away, but I was wondering if it were still lurking around. I finished the exposure. I waited about three-quarters of an hour before I went out because I wanted it to get light. Before I closed the door, I looked all around. When it looked like the coast was clear, I dashed to the administration building.[15]

Although he was never sure, Tombaugh concluded that the animal must have been a mountain lion. Two years later Kenneth Newman had a frightening experience one very dark, starlit night while helping Lampland with work at the 42-inch; Tombaugh was nearby taking planet-search plates. Newman was going back to the administration building to get more plates for the large reflector when he noticed an animal moving slowly across the ground. Thinking it was the observatory dog, he moved toward it and bent down to pet it; the animal backed away. The following morning, machinist Stanley Sykes went to the 42-inch and saw the full story written in tracks in the snow. A human had gone out of his way to approach a set of animal tracks. "Someone got really close to a mountain lion last night," Sykes told the astonished Newman.

During the June lunation V. M. Slipher showed Tombaugh a fine 5-inch camera, with a focal length of 22.35 inches, that the observatory had borrowed from Wilbur A. Cogshall at Indiana University. "I'd like for you to run this camera also, for confirmation," Slipher explained. The instrument used a Cooke triplet lens. The lens was attached to the 13-inch on June 7[16] so that any planet suspects fifteenth magnitude or brighter could be checked. About the same size of instrument Lowell had used for his first search, the camera produced 8-by-10-inch plates that were not generally used for blinking. On these plates one angular degree covered precisely one centimeter. The larger scale of the 13-inch, close to three centimeters per degree, allowed much more complete searching.

Even though this camera produced beautiful plates, Tombaugh

"found it more of a nuisance than anything else because most suspects were beyond its reach." It virtually doubled the amount of work he had to do in preparing, loading, unloading, and processing plates. Starting and ending exposures were simplified after he devised a mechanism that opened both 13- and 5-inch shutters at the same time using strings that were adjusted to move them the right amount. Even the suspects that did appear on the Cogshall camera plates had to be checked with a third plate taken another night. Nevertheless, the Cogshall project was worth the effort in one respect. An atlas of the entire sky visible from Flagstaff is preserved on a convenient scale with generous overlap, since these plates show much more sky than the plates taken with the 13-inch instrument.

The June lunation ended three months and about a hundred plates of photographic work, during which Tombaugh had progressed from Gemini all the way to Scorpius and Sagittarius. These plates, taken from April 6 to June 17, ranged from the relatively star-poor regions of Cancer and Leo to rich fields in the Milky Way. The Sagittarius plates especially were blackened with the dense star fields. Immediately after the Gemini plates had been processed, V. M. and E. C. Slipher blinked them in the hope that their "fast find" approach would reveal Planet X. After almost a week of fruitless work, they stopped. No planet had turned up. Later Tombaugh recalled that the plates were taken as a possible trans-Neptunian planet would have just passed the western stationary point, meaning that the object would hardly have been moving at all from one exposure to the other. Convinced that no one at Lowell realized that at the time, he later saw that this early blinking effort was doomed before it began.

Since the Moon was now bright, and the prospect for clear weather in July was diminished by the predicted onset of summer storms, Tombaugh looked forward to a slackening in his schedule, a chance to catch up on his reading. On June 18 all that changed. V. M. Slipher entered Tombaugh's office to say that the blinking would from now on be his responsibility. For a moment Tombaugh just stared at his boss. How could this be? Tombaugh had examined his plates superficially, assuming that the "heavy responsibility of finding, or not finding, the planet" would fall on more experienced people.[17] This unexpected change in plans could mean but one of two things: that the men now had enough confidence in Tombaugh's ability and discipline to do the job, or that they had concluded from their earlier blinking that Planet X was probably not there anyway and that the search should be given lower priority. Were they already giving up? The answer was probably a combination of both reasons.

Tombaugh's Blinking Begins

Two pairs of plates were all it took to discourage any conscientious searcher. Planet suspects appeared everywhere; how could one say that any of them was not the long-sought Planet X? As the spring went on, two problems became apparent. Everyone was uncertain how to distinguish between a near asteroid and a distant planet, and no one was blinking plates. Lampland, who had returned from a semester's teaching at Princeton, was so inundated with work that he could not blink for a while, a most unfortunate circumstance since he was far more familiar with the blinking process than anyone else on the staff. Some hundred plates had been taken by the end of June, and if the missing planet was on two of them, who would know? Two of the early plates had been centered on Delta Geminorum and did indeed contain images of Pluto. In their hasty search of these plates one of the Sliphers missed the images, probably because Pluto, being so far from opposition, was nearly stationary, hardly moving among the stars, and thus showing very little apparent motion from one plate to another.

With Planet X believed to be lurking somewhere in Gemini, Tombaugh dreaded the prospect of looking through those huge numbers of stars, estimated in that region to be about three hundred thousand per plate.[18] "I was almost soured with it. The thing that disturbed me was that Lowell had changed his prediction [in Libra] several times. Was the prediction so shaky that it could make that big of a change?"

Although the renewed planet search was not generally publicized, word did leak out. Throughout much of the astronomical community the reaction was negative, as it was with most of the other projects Lowell and his observatory had undertaken. The Martian canals had been a serious (and in retrospect unnecessary) embarrassment to the observatory, and although other people had been involved in mathematical searches for a trans-Neptunian planet, so many years had passed without any success that astronomers generally considered it a hopeless cause. Lowell himself had suggested that his associates engage in other projects far removed from the solar system in order to gain credibility for themselves and the observatory.

Despite the new and wonderful telescope, the prospect for a successful conclusion to the search for Planet X appeared dismal. With Tombaugh now saddled with much of the responsibility for this search, the uncertainty plagued him.[19] The rest of June left him exhausted and depressed; having blinked two pairs of plates, he had encountered many objects for which there was no way of proving whether they were asteroids or distant major planets. Plates had anywhere from just a few to more than one hundred moving objects on them.[20] He felt that he had been saddled with a dead-end project. Moreover, patrol work in astronomy is inherently a risky business. An observer can spend a lifetime without any success at all, without publications, without even a strong suspect. Worse, the observer can become careless and report a suspect that turns out to be an asteroid or even an emulsion flaw in the plate.

Adding to Tombaugh's dejection at the time was the possibility that even if he did stumble onto Planet X, he would have no idea how to confirm it. Surely he could rephotograph the suspect, following it for a while, but since he was seeing several such suspects on every plate, the work of following up every one of the suspects would be prohibitive. In any event, Tombaugh's unhappiness was compounded by simple homesickness, symptoms of which Lampland noticed before anyone else. Probably Lampland suggested to V. M. Slipher that a vacation trip home would cheer Tombaugh up a bit. As the train pulled out of Flagstaff and Tombaugh saw the now-familiar mountain scenery receding, he wondered if the purpose for his being there would ever be fulfilled. He left behind three astronomers concerned that he would abandon the project. "They were under the gun to get the job done, and I think they were scared that I wouldn't go through with it."

Autumn 1929

The real thinking commenced after Tombaugh returned from vacation. How could one differentiate between a trans-Neptunian planet and an asteroid near its stationary point? As the outline for a solution seemed to form in his mind, he compared the apparent motion of the outer major planets over three years from the *American Ephemeris and Nautical Almanac*.[21]

The major planets orbit the Sun at different distances. Since Mercury and Venus are closer to the Sun than Earth, they are seen approaching the Sun, then receding from it as they wander through the sky. Mars, being farther from the Sun than Earth, will apparently travel completely through the zodiac, being close to the Sun at times, but at other times being in opposition.

Why not take all the plates at this opposition point? With that strategy any solar system object found there will be retrograding at a speed in reverse proportion to its distance from the Sun; for an object with a nearly circular orbit, the greater the distance, the smaller the daily shift. "As a consequence," Tombaugh later wrote, "the asteroids, on the average, moved about seven millimeters per day on the plates, and exhibited short trails during the hour's time of exposure, whereas Pluto moved only a half millimeter per day."[22] If the plates are taken at opposition, the motion of an object (particularly one with a circular orbit) would be a function of its distance from Earth. Once the rates of asteroids and the outer planets Mars through Neptune were known, it would be possible to use the retrograde motions themselves as the identifying key. The shorter the motion, the farther the object; it was that simple.

The best way to think about this apparent motion of a planet is to imagine what happens when you drive a car on a highway. A car ahead appears to move forward more slowly than yours. As you set out to pass the other car, it appears to slow down, and if you watch it as you are overtaking it, it seems to be moving backward. The eastward motion of Mars in the sky is similar. Approaching the opposition point, it appears to slow down and for a short time is almost motionless. This is called its *stationary point*. Then, as Earth overtakes it, Mars appears to move westward, or in retrograde. After a while the speed of westward motion decreases, another stationary point is reached, and then eastward motion resumes until the next opposition. The asteroids in the main belt, as well as the other outer planets, all display the same behavior at opposition. There is one important difference: the farther away a planet is, the more slowly it moves when it is at opposition.

That was the solution. Now there was a way.

Observing Procedure

The stock market crash in October 1929 left most of the world in a confused and worried state. Within a few months anyone who had a job at all would be fortunate. Tombaugh, however, was having the time of his life. When he started, he was making $90 per month, plus the use of an unheated room on the second floor of Lowell's administration building. After about two months Slipher raised his salary to $100, later to $110, and finally to $125.

Anyone who searches for faint solar system objects leads a life dominated by lunations or moondarks. The full Moon period, lasting a few days during which little dark sky is visible, begins a lunation. As it wanes, the Moon rises later and later each night, revealing a longer period of dark sky. The new Moon period lasts a few days during which the sky is completely dark most of the night. Professional observatories schedule their telescopes around certain times of each lunation; *bright time* occurs near full Moon, *dark time* near new, and *gray time* around last and first quarters. The 13-inch was not concerned with such allotment of time, since it was blessed with the astronomer's dream of being completely dedicated to the single project of the planet search. The times during each lunation required careful planning, however, with regions getting away from the opposition point being photographed during the first nights and other areas coming later.

On a clear evening Tombaugh would first "walk around" the instrument, turning on the drive, opening and closing the shutter, moving the dome back and forth, to make sure that everything felt and sounded right. He might then find a star on the meridian and look at it through the guiding telescope's high-power eyepiece, hoping that the image he saw would be steady, not, as astronomer Robert Burnham later described, "trembling like a piece of Jell-O."[23] Poor seeing would mean "soft" and swollen images that were affected by turbulent air and took up too much room on the plate, resulting in the faintest objects not appearing at all. A light haze is preferable to a clear sky with poor seeing, because it does not cause star images to swell and can be compensated for by increasing the exposure. Even if the seeing at the start of the night were poor, he would still continue by loading the plate holder, since the sky could improve later. He would wait until the first plate's region was high enough and then, using the setting circles, turn the telescope to the correct right ascension and declination of a star near the region's center which he had chosen as a guide star. Focusing the star in the guiding telescope, he would then recheck the seeing, check the time, and open the shutter.

During the next hour the telescope would move one twenty-fourth of the way around the sky. Although the weight-driven telescope tracked the star very nicely, Tombaugh kept his eye at the eyepiece much of the time just to make sure the guide star was right at the center of the field. Occasionally, he would move the right ascension just a bit, and the telescope would respond. Periodic looks at the time would monitor the exposure's progress. At the end of the hour, or longer if the sky was hazy, Tombaugh would close the shutters and get up from his chair, stretching a bit. Quickly, he would remove the plates and store them, complete the recording in the observing log, and load a new set of plates for the next exposure. The procedure would be repeated until the Moon rose or until the opposition point got low enough that bad seeing would prevent further useful work.

This procedure worked well on a clear night, but most nights are hardly ideal. Patchy cirrus clouds would cross the sky for days on end, and even after storms, cumulus clouds would interfere. On these nights Tombaugh would have to check the sky frequently to see if an exposure could begin or, if one was already in progress, whether it would have to end early. For a passing cumulus cloud Tombaugh could simply close the shutter until the offender was gone, and then he could reopen the shutter and continue the exposure. If the presence of cirrus clouds was the problem, the decision to continue or stop was often a tough one. On such nights Tombaugh would work even harder than on clear nights, for he would constantly look up from the slit in the dome to check the clouds as well as checking the guide star's position in the eyepiece; in any case the results would rarely be as good as work done on a completely clear night. Deciding whether to observe on such nights would depend partly on the severity of the cloud problem and also on how far behind Tombaugh was in his work. A badly needed second or third plate would keep him anxiously looking at the clouds most of a night. If he could succeed in rephotographing an area, he could leave it off the next moondark's observing plan. Only rarely did Tombaugh have to stay up an entire night. Usually by three in the morning the opposition point would have sunk too low in the west to permit good exposures. Atmospheric seeing deteriorates rapidly the lower in the sky a region gets.

Unless the sky was hopelessly cloudy at the start of a night planned for observing, Tombaugh would have a plate loaded and ready to go. He would then wait, usually in the administration building, spending the time reading by a soft light that would help to maintain his dark-adapted eyes. Frequent interruptions would occur as he walked outside to check the sky. After his encounter with the mountain lion, during these nights he would not normally walk around outdoors more than was necessary to check the clouds and walk to and from the 13-inch. If the sky showed substantial signs of clearing so that increasingly large open patches were appearing, particularly from the west (or southeast in summer), he would then move to the telescope and completely dark adapt. From his list of priority areas he would choose the one best placed relative to the clouds. If the clear patch got large enough, he would center the chosen star for that region in the 7-inch guide scope that was attached to the main instrument. If the sky continued to improve, he would start the exposure.

To pass the time while guiding an hour-long exposure, Tombaugh would continue the little mental games he had been so fond of on the farm, such as calculating the number of cubic inches in Betelgeuse. Now he would design telescope systems in his head, solving the problems of the astronomical world during an exposure.

At the end of a long night Tombaugh would close the telescope down, turn off the drive, cover the optics, shut the dome, close the front door, return to the administration building, store the plates in the darkroom, and finally crawl into bed. There are few feelings more pleasant for an observer than putting head to pillow after a long and fruitful night. Even after long winter nights Tombaugh would rarely sleep much beyond eleven the next morning. A check of the sky (automatic for any observer) and a light meal would start his day. If the sky looked clear and he was in midlunation, he would not consider doing much blinking. Instead, he would conserve his energy by developing plates and planning for the night ahead.

On the nights around new Moon, with clear sky, Tombaugh might put in fourteen-hour or longer days. A long string of clear nights around new Moon is not as common as might be hoped, however. Under no conditions could planet-search plates be taken with the Moon in the sky when it was more than three days from new. Even a crescent Moon's light would completely darken the emulsion, washing out stars and making comparison with another plate impossible.

This careful procedure for an observational planet search was a very different thing from the quick find that produced Neptune. In 1846 Galle and d'Arrest simply moved their telescope to the position Leverrier had sent them and within an hour found the interloping planet. The search for Planet X had not gone so easily. It seemed obvious that it would not be found precisely at a position determined mathematically and that a much wider search was needed, a search the Sliphers hoped would not take more than a few months but that they feared might last years. Although the renewed work began far to the west of Lowell's predicted place, as the warm early fall of 1929 turned cold in November, V. M. Slipher asked Tombaugh often if he was finding any suspects.

During the first blinking session in September, Tombaugh examined a region in Pisces. Suddenly, a huge black spot appeared and disappeared on one plate, then did the same thing on another. Uranus! He had known that William Herschel's planet was somewhere around but had deliberately not checked in the hope that he would bump into it accidentally. "Being of the sixth magnitude, it was a real wallop. I almost ducked my head."[24]

One afternoon early in the fall of 1929 E. C. Slipher, possibly to offer some encouragement, showed Tombaugh a plate taken long ago on which an object had been specially marked, and explained that this could be the object of their search. Astonished that they hadn't tracked it down with a third plate, Tombaugh stared at Slipher. "You mean you don't really know?" With this admission, Tombaugh knew that Planet X could now be found if it existed and were bright enough. What unpredicted objects could turn up as well? On that afternoon Tombaugh saw the sky open up for him. He felt he could search the whole zodiac, whether planets were predicted or not. So long as he was careful, a planet could be found. If the others had not even bothered to check out that mystery suspect, what other objects might be there? On that afternoon Tombaugh's feeling was confirmed; the project in a sense passed out of the mathematics of Percival Lowell and into the observational technique of Clyde Tombaugh.

When the observing season began in the early fall, Tombaugh was a new man. He had figured out the "X" problem, and in the following years he would find the planet or determine that it was either not there or fainter than the magnitude limit on the 13-inch plates. The entire project was his, and he was approaching it with pride. That worked out very well for V. M. Slipher as well. "He had sacrificed months of work to get this telescope set up," Tombaugh recalls, "and he had really done a good job. So I think he appreciated the fact that I had taken hold of it so thoroughly."

The 24-inch

Quite often on moonlit nights when Tombaugh was not taking search plates, E. C. Slipher would replace the camera with the eyepiece on the 24-inch refractor and invite him to observe Mars during plate processing. Tombaugh enjoyed this telescope immensely, even more when it provided a break from the routine of blinking. E. C. Slipher was also proficient at visual observing. "Oh, how that man could draw!" Tombaugh exulted. "I never saw any published sketches that beat his." [25] To be skilled at both visual and photographic observing is rare indeed. "He'd invite me to come up and take the camera off the 24-inch, and I'd look at Mars and the other planets visually. I just loved that." On other occasions V. M. Slipher let him guide a spectrograph for the radial velocity study, and once he and Tombaugh took a spectrogram of Jupiter's south equatorial belt.

Since planets often reveal subtle details when filters are used, a set was usually mounted in a rotary holder on the 24-inch for easy access. Although many observers prefer orange or red filters, E. C. Slipher favored the yellow filter for Mars. Through the refractor the light at the violet end tended to be most annoyingly defocused, and this filter blocked most of that light. Tombaugh also liked yellow because it did not alter objectionably the natural color of the planet.

The large refractor telescope had a diaphragm that could stop the lens down to as small as 6 inches, and on all but the best nights the aperture would be stopped down to 16 inches. Looking at the observatory founder's old records, Tombaugh found that Lowell himself had stopped the instrument down to 16 inches almost 90 percent of the time. Stopping down a little more would help on poorer nights. Tombaugh noticed a dramatic improvement on one unusual night when he simply stopped the diaphragm down from 16 to 14 inches. Our atmosphere contains "seeing cells" of certain diameters that are approximately defined, although on that night the drastic improvement indicated that they were quite sharply edged. Tombaugh watched many oppositions of Mars with Lowell's great refractor. Over these years he saw many of the "canals" that Lowell had seen: "I've got lots of canals in my own drawings," Tombaugh explains, referring to the 125 canal-like features he observed at one time or another, adding that unless the seeing is nearly perfect, an observer won't see them at all. With even better seeing and the McDonald Observatory's 82-inch telescope, Tombaugh later saw the canals broken into the smaller details that spacecraft have shown them to be (see Chapter 10).

Another problem with the 24-inch was that the color aberration prevented an observer from seeing more detail because he couldn't use high enough magnification. Beyond about 500 power the images would deteriorate. During his years at Lowell, Tombaugh argued with the older men about the merits of refractors versus reflectors. Tombaugh felt that a high-quality reflector would have images just as good as those of a comparable refractor, but without the color problem. Based mostly on their experience with the imperfect 42-inch reflector, however, the others preferred refractors. "I always had this little contention with them that a reflector properly made with a good focal ratio would equal or surpass a refractor. But I never could convince them of that. When I got here [Las Cruces] to my own empire, I got the 24-inch reflecting system, and you see how well we have done" (see Chapter 12).

For as long as Tombaugh can remember, a 12-inch refractor was mounted on the east side of the 24-inch. Added some time after the larger telescope was completed, it was used to monitor the seeing during photographic work with the large telescope. The idea was that when the seeing settled down as seen through the 12-inch, an exposure would be taken through the 24.

The big refractor offered only leisure-time observing for Tombaugh, who was now headlong into the search-and-discover mission for Planet X. During the September lunation he concentrated on Aquarius and Pisces, and in October he moved into Aries. November's run saw the start of Taurus, and in December he was well into the eastern part of that constellation. By January 1930 Tombaugh had reached the region in Gemini he had photographed eight months earlier. But now the area was in opposition, and since it was near Lowell's favored position, it would have to be rephotographed.

January 1930

On January 21 Tombaugh began an hour exposure guided on Delta Geminorum. The first few minutes were uneventful, but then a biting wind started from the northeast, rapidly increasing in strength. The seeing deteriorated rapidly, Delta Geminorum swelling almost beyond visibility in the guiding eyepiece. The exposure was repeated two nights later, and another good plate was taken on January 29.

As he blinked the plates of a region in eastern Taurus, Tombaugh slowed down rapidly when they became crowded with huge numbers of Milky Way stars. He was getting further and further behind in the blinking part of his operation and had not even reached the densest part of the Milky Way. To lessen the number of stars he would see at one time in the Milky Way, he would put a diaphragm into the comparator's eyepiece that would allow only a narrow swath three millimeters wide and twenty long. Without the diaphragm he would get so confused by the many stars that he might not be able to tell where he had left off.

If he stayed in Taurus he would fall so far behind that if he were to find Planet X, the region might be too low to rephotograph. The thought of skipping over for the time being the two sets of plates that covered the densest Milky Way regions occurred to him. He had already photographed these areas; the plates would simply wait until a more convenient time for blinking. The Delta Geminorum plates on the other side of the Milky Way beckoned with far fewer stars, and Tombaugh could search the same area of sky in two to three weeks, compared to four or five weeks, at a rate of a few square inches per day, for the richest Milky Way regions. Since planets could be anywhere in the zodiac, searching in a region poorer in stars would be more efficient. Since Gemini was the region Lowell had most favored anyway, the plan would definitely meet with V. M. Slipher's approval. Maybe the new strategy of searching at opposition would work where the quick-find approach had failed. Accordingly, in mid-February he placed two Delta Geminorum plates, one taken on January 23 and the other on January 29, into the comparator.

Chapter 5

DISCOVERY'S WAKE

You have to have the imagination to recognize a discovery when you make one. When they examined Voyager images and saw for the first time the volcanic eruptions on Io, that called for some intuitive imagination. I would suggest that above everything else, in observing you have to be very alert to everything. You have to be able to recognize a discovery as such. There are so many people who don't seem to have that talent. A research astronomer cannot afford to be in such a rut. I might say that different types of personalities in astronomy make certain types of discoveries that are in line with their personalities.

—Clyde Tombaugh, 1985

TOMBAUGH'S comment recalls a number of types of discoveries. The second and third discoverers of Comet Kobayashi-Berger-Milon, for example, were separately observing the globular star cluster Messier 2 when they stumbled on the unfamiliar object in 1975. Each recognized it and reported it as a comet. Dennis Milon was an experienced comet observer, and Douglas Berger a talented watcher of deep sky objects. Neither was hunting for comets at the time, but both had the expertise to know what to do.[1] Other observers were slower to recognize the new comet near Messier 2 or for other reasons did not communicate their find in time. Although the criteria for discovery of a comet—shape, appearance, and motion with respect to star background—are relatively straightforward, the observer needs to know what to look for.

The discovery of bright Supernova 1987A in the Large Magellanic

Cloud was reported by three observers. In Chile, when Ian Shelton's search plate for novae was processed, Shelton immediately noticed a bright object, went outside, and observed it visually. On the same mountaintop Oscar Duhalde, an observing assistant, noticed the interloper but did not report it until after Shelton did. Even then there was delay in figuring out how to send the report. In the meantime, Albert Jones, one of the world's most experienced variable star observers, discovered the supernova from Australia hours later. Because he recognized it at once for what it was and knew how to report it, his discovery reached the Central Bureau for Astronomical Telegrams almost at the same time as the report from Chile.[2]

Finding supernovae in distant galaxies should be a straightforward matter, requiring a thorough knowledge of what many galaxies look like and the intuition to recognize a faint speck of light that arises inside one as an interloper. Nonetheless, few people have the imagination and perseverance that is needed to make repeated discoveries, and only one, the Reverend Robert Evans, has carried his program through to make eighteen visual discoveries in fewer than ten years. Evans and Tombaugh share the dream to discover, the patience to develop the procedure, and the perseverance to continue the search until it succeeds.

This question of the kind of person who makes discoveries came up over and over again during the author's interviews with Clyde and Patsy Tombaugh. Tombaugh cited the case of a university whose vast archives and research concentrated on variable stars. "If they had found Pluto there, they might have said that it was simply two variable stars out of phase with each other!" Of course, a thorough investigator would surely have checked other plates and not noticed either variable star. It would have then taken some imagination to figure out that the objects were not an asteroid, but a far more distant solar system object.

This discussion does not imply that astronomers who miss discoveries are not good astronomers, only that those with additional imagination have a better chance at finding unexpected objects. For example, James Christy discovered Pluto's moon, Charon, in 1978, as an appendage that could well have been mistaken for a blemish on Pluto's image. That discovery demonstrated conclusively that Pluto had far too little mass to have been the cause of the orbital perturbations of Uranus and Neptune and finally showed that Pluto's orbital similarity to Lowell's

Planet X was but a happy accident. It also meant that it was the observational search, not the mathematical one, that led directly to Pluto's discovery. "I was not impressed with Lowell's prediction," Tombaugh now admits. "He had a prediction in Libra, then he changed it to eastern Taurus, and then Gemini. If it isn't better than that, how good is any of it?" Although Tombaugh never expressed his feelings to the Sliphers, he did not have much confidence in any of the mathematical predictions, including Lowell's. It was only because Pluto's position agreed so closely with Lowell's prediction that many astronomers, including Tombaugh, believed it was likely Lowell's Planet X.

The days after the discovery were thrilling and tense. With the eventual announcement, Clyde Tombaugh gained fame around the world, at the cost of losing some of his private life. The discovery was the climax of a long search sired by Lowell, encouraged by Putnam, organized by V. M. Slipher, assisted by E. C. Slipher and Carl Lampland, and concluded by Tombaugh.

February 19, 1930

Not many hours after he had gone to bed, Tombaugh awakened. Just for a moment he turned sleepily around, his eyes closing again. Planet X! Instantly, he felt better than he had in all his young life; the pleasure swept over him in waves. Out of bed in his little room he looked outside. The clouds were still there, but the sky looked a bit brighter. Maybe it would clear tonight for a new plate of the area around Delta Geminorum.

After breakfast Slipher, Lampland, and Tombaugh sat down to discuss strategy. The tension was very high that morning; "everybody was as excited as the dickens!" It was decided that Tombaugh would, for the time being, continue to photograph the Delta Geminorum region and that Lampland would get plates with the 42-inch reflector so that more precise positions would be measured. Although it was a little early, some thought was given to working out an orbit. During much of this day Tombaugh examined the "X plates industriously, checking up on new object—on plates made last year."[3] Apparently, E. C. Slipher was telephoned on February 19 or 20 in Phoenix.

On the night of the 19th Tombaugh went to town for dinner much earlier than usual. By nightfall there were still some clouds, but the sky looked good enough to set up. Despite occasional mist, he went out around eleven o'clock and succeeded in getting his plate.[4] He processed it at once and rushed it to the comparator room. Yes, it did show the distant traveler precisely where it was supposed to be, a bit to the west of its last position. At Lampland's suggestion, he then made a small contact film, about five inches in diameter, of the area immediately around the object.

The Next Days

During the following days the small staff was busy considering two major aspects of the discovery, the first being attempts to photograph the object again and the second involving the question of when to announce it. The reaction of everyone was to proceed cautiously, watching the planet's progress among the stars for some weeks. The object's faintness made accidental discovery by someone else unlikely. Since Percival Lowell's canal controversy, the observatory had had a sort of underdog status. Clearly, the object the staff members now knew about represented their best chance to regain their reputation as a first-class research institution.

By February 20 it was time to examine the new object visually. With his contact print of the previous night's image to use as a finder chart, Tombaugh walked to the 24-inch refractor with Carl Lampland; V. M. Slipher joined them as they walked.[5] They opened the dome, removed the covers of the long refractor that for so many years had gazed on a sky with eight planets, and turned the telescope toward the ninth. Carefully, they studied the object, Tombaugh feeling particularly moved that for the first time human eyes were gazing on it.

The look was a bit disappointing. Even at the best magnification the planet revealed not the slightest hint of a disk; could it be that small? At first they had hoped that its extreme faintness could be explained by a large object with low albedo, or reflectivity. Without a perceptible disk, there would be doubts that this object had sufficient gravitational pull to affect Uranus and Neptune and be Lowell's mathematically computed Planet X. The time spent taking the visual observation meant that Tombaugh took no plate that night, although he did on the nights of the 21st and 22nd.

On the 23rd V. M. Slipher finally wrote to trustee Putnam. It was a most unusual letter, beginning with a casual discussion about observatory and world affairs, following with a statement that Mr. Tombaugh had found a "very interesting object" that may be Lowell's Planet X, and ending with comments about Mars.[6] Quickly losing interest in the stock market and Mars, Putnam was, needless to say, absolutely delighted with the news.

Not long after the discovery a thought began to prey on Tombaugh's mind. Thinking that the search was over and the tedious blinking was done, he had stopped blinking immediately after discovering the new object; he did not go a single field further. In the meantime, with the 42-inch, Lampland had begun a search for possible satellites of the new object. What if the object Tombaugh had found were the satellite, and the main planet were somewhere nearby, much larger and brighter? The idea began to trouble Tombaugh within a week of the discovery, after the visual attempt to see a disk had failed. By this time the staff members were under strain; under increasing pressure to make an early announcement, they worried about the publicity about to descend on the quiet observatory. To all this tension Tombaugh didn't want to add his own fear that he may have found only the satellite of a larger body.

Unfortunately, Lampland had replaced both of Tombaugh's plates with his own plates taken with the 42-inch reflector. He also had removed the comparator eyepiece, replacing it with a much higher-power and delicate micrometer eyepiece, so he could measure the object's positions in preparation for an orbit computation. Technically, the comparator was unavailable and would be for two months. It would not have taken more than an hour to remove Lampland's plates, change the eyepiece, put the discovery images back on the comparator and blink further, and then replace everything. Although he realized that searching just a few centimeters farther would have answered his question and no doubt allayed his fear, Tombaugh never did it, instead spending more than a week fearing that Lampland, or worse, someone from another institution, would find the larger object. "That just scared the living wits out of me for a while."

When the author asked why Tombaugh never replaced the plates and conducted the brief search, he replied that he would have been somewhat embarrassed to admit this possible oversight. Moreover, he didn't think to do that work surreptitiously, perhaps late at night when the others were asleep. It was not in Tombaugh's nature, one supposes, to be covert, so he kept his thoughts to himself, and his worry faded after Lampland had searched the 42-inch plates and neither satellite nor larger object turned up. "I could have committed a terrible blunder. Boy, would my face have been red!"

Although Slipher had suggested that Tombaugh take some further images of the Delta Geminorum region when the Moon finally left the evening sky and the next lunation was underway, Tombaugh felt that there was no reason to do so. Lampland was taking images using the 42-inch reflector virtually every clear night, and the smaller plates were much less expensive and their scale was more appropriate for the astrometry, the determination of precise positions, that was needed. Lampland was also using these plates to compare the color of the new object with that of Neptune, and he confirmed that the object was yellowish, whereas Neptune was bluish.[7]

The constellation of Cancer was now passing the opposition point. Why not simply keep up the observing schedule as though nothing had happened? After taking some nights of further Delta Geminorum plates, Tombaugh began following the zodiac again, taking search plates of the Cancer and Leo regions. They would not be examined right away, but at least the search program would be kept up. At this time the other astronomers had given no thought to continuing the search. Even at this early date, Tombaugh felt that continuing the photographic program was a possibility; in any event, it seemed the logical thing to do.

Soon after the visual observation, E. C. Slipher returned from Phoenix and came up with a most interesting idea for an experiment. Hoping to determine how large a disk the object would have and still have the disk escape detection, he mounted a box punctured with holes of different sizes and light levels on the roof of a Flagstaff hotel. Because of the object's faintness, Slipher's results indicated that if its disk were larger than half an arc second, at its fifteenth magnitude brightness such a disk could not escape visual detection with the 24-inch refractor.[8] Although this experiment did give an indication of the size of the disk, it was not perfect in that the path was always close to the Earth's surface.

Tombaugh spent much of this period conducting his usual duties, including shoveling snow, loading wood and coal into the furnace, and, according to Lampland, other assorted "janitor work." The Sliphers and Lampland also spent much time vainly trying to find images of "X" on the plates taken by Tombaugh in 1929. Lampland wrote, "V. M. Slipher looked further at last year's plates, I understand from Tombaugh that is the outstanding cause of doubt as to the reality of the object being 'X'. Of course one might assume at least two possible causes to account for the absence but these are rather remotely possible."[9] These two causes are a poor predicted position because of the lack of a good orbit and because of the fact that the earlier pictures were taken far from opposition.

As the days quickly passed, attention began to focus on the announcement. Trustee Putnam urged V. M. Slipher to announce as soon as possible, as did Harlow Shapley, who was anxious to publish the news in his *Harvard Announcement Cards*. The question now was whether to watch a little longer, perhaps two weeks, or to announce right away. A unique date, March 13, was fortuitously appearing on the horizon: it was both Lowell's birthday and the 149th anniversary of the discovery of Uranus. If the planet continued on its course until then, the announcement could coincide with this date.

Arguing in favor of a more rapid announcement was the fact that Delta Geminorum was already well up in the sky at sunset, setting almost four minutes earlier each night. Astronomers would want to follow the object at once, and if the observatory waited, the object would soon be setting too early. In favor of caution, as V. M. Slipher recalled from Percival Lowell's Martian canal controversy, was the memory of how embarrassing a premature and inaccurate broadcast could be. Normally, discoveries of comets, asteroids, novae, and supernovae are announced within days or hours, as soon as the discoverer is certain of the status. But the object in this case was a major planet, and one can surely understand V. M.'s desire to watch the motion for a short while to make sure that it conformed to that of a trans-Neptunian object.

Before making the announcement Slipher also wanted to try to get a better understanding of the object's distance than he had thus far simply by the amount of the shift at discovery. The staff knew about when the object should start to slow down as it approached its stationary point; if it was truly trans-Neptunian, it would reach this point later and more slowly than would an object at Neptune's distance. On March 2 the object was continuing to move as expected, "about 44″ per day."[10] As March 13 neared, Slipher became aware of another reason for delaying the announcement. The staff members were becoming increasingly tense as they prepared for the storm. The extra time, Slipher felt, was important for observatory morale.

Unfortunately, a question of credit was also raised. Tombaugh recalls a conversation he overheard between V. M. Slipher and Carl Lampland. Tombaugh was in his office and did not hear all that was said, but apparently the voices became loud enough that he heard Lampland's remark from down the hall: "Well, he is the one that did the work. He did the blinking; he's the one that found it!"[11] As it turned out, Tombaugh was not mentioned at all in the telegram, but he was in the Lowell Observatory circular.

Tombaugh was the last person to want to rush things. The weeks between the discovery and the announcement were exciting, but relatively quiet. V. M. Slipher warned him to be careful about what he would say to the press and to watch out for people who might try to get benefit for themselves from the discovery. Tombaugh had moved from a quiet farm life to the observatory just over a year earlier; soon he would be a very public person, and the direction of his life would show a gargantuan change. He was enjoying these last peaceful days while he could: "I thought that from then on there would be a lot of excitement and pressure. And in some way I dreaded it a little bit. I was naive and didn't know the ways of the world, and Slipher wanted to caution me against getting into a jam. It was very kind of him to do that." Was it possible that one of Slipher's ways of protecting Tombaugh was to deny him full credit for the discovery?

On March 8 a long period of uncertain weather ended and the sky cleared. Lampland's diary recorded "beautiful morning" (March 9); "Spring weather" (March 10); "Glorious spring day" (March 11); "Most delightful weather" (March 12) as announcement day approached. By March 11, V. M. was making drafts of the announcement, while Tombaugh was measuring the object's magnitude at 15.25.[12]

The Announcement

Just after midnight on Thursday, March 13, Tombaugh joined V. M. Slipher in secretary Constance Brown's office as he sent the following telegram to Harlow Shapley at Harvard, via Roger Putnam: "Systematic search begun years ago supplementing Lowell's investigation for Trans-Neptunian planet has revealed object which since seven weeks has in rate of motion and path consistently conformed to Trans-Neptunian body at approximate distance he assigned. Fifteenth magnitude. Position March twelve days three hours GMT was seven seconds of time West from Delta Geminorum, agreeing with Lowell's predicted longitude." [13] Shapley immediately published *Harvard Announcement Card* No. 108 and informed the International Astronomical Union's Central Bureau in Copenhagen for distribution to the Eastern Hemisphere.

After sending the telegram, Slipher turned his attention to an observation circular entitled, "The Discovery of a Solar System Body Apparently Trans-Neptunian," a publication Tombaugh had helped prepare in two ways, first by discovering the planet to which it referred and second by helping Constance Brown stuff hundreds of them into envelopes. Tombaugh remembers having the time of his life folding all those circulars.

The circular began by quoting the telegram and then gave a right ascension of 7 hours 15 minutes, 50 seconds of time, and a declination of 22 degrees, 06 minutes, 49 seconds of arc. One wonders why the declination was published to greater accuracy than the right ascension, since a second of time is fifteen times larger than a second of arc, and why the equinox of the observation was not included, as positions need to be referenced to a coordinate system prepared to reflect a specific year, such as 1855, 1900, or 1930. Traditionally, positions given without such a reference "equinox" refer to the current positions as seen with the telescope, but an announcement as important as this one should have been more clear. All Tombaugh's unpublished plate notes, for instance, have the guide stars listed to appropriate precision in right ascension and declination, and most specify equinox 1930. In view of the three and a half weeks the staff had to prepare this announcement, as well as the observatory's later reluctance to reveal *any* positions, perhaps these omissions were deliberate. The circular reads:

The finding of this object was a direct result of the search program set going in 1905 by Dr. Lowell in conjunction with his theoretical work on the dynamical evidence of a planet beyond Neptune (See *L. O. Memoirs*, Vol. I, No. 1, "A Trans-Neptunian

Planet," 1914). The earlier searching work, laborious and uncertain because of the less efficient instrumental means, could be resumed much more effectively early last year with the very efficient new Lawrence Lowell telescope specially designed for this particular problem. Some weeks ago, on plates he made with this instrument, Mr. C. W. Tombaugh, assistant on the staff, using the Blink Comparator, found a very exceptional object, which since has been studied carefully. It has been photographed regularly by Astronomer Lampland with the 42-inch reflector, and also observed visually by Astronomer E. C. Slipher and the writer with the large refractor.[14]

The circular continued with detailed information, including a carrot dangled for astronomers hungry for precise positions they could use in calculating an orbit: "Besides the numerous plates of it with the new photographic telescope, the object has been recorded on more than a score of plates with the large reflector." It ends with a statement of concern that it has been announced "before its status is fully demonstrated; yet it has appeared a clear duty to science to make its existence known in time to permit other astronomers to observe it while in favorable position before it falls too low in the evening sky for effective observation."[15] In discussing this announcement with Tombaugh, the author learned that he was not altogether happy with the wording; there was too much information revealed about certain items, such as the apparent size and the numbers of plates Lampland took, but too little about other matters, such as a description of the blinking procedure. Tombaugh felt at the time that his junior status prevented him from expressing his feelings.

Recalling the sensation over Uranus and the controversy over Neptune, the staff at Lowell was not sure what to expect after the announcement of the first new major planet in eighty-four years. The first unpromising hint came at Arizona State Teachers College on the afternoon of Thursday, March 13. As it was Lowell's birthday, the annual presentation of the Lowell Prize in mathematics had to be made at the school, later Northern Arizona University, which was taking its place as one of Arizona's three main institutions of higher learning. Usually it was the director's task, but the discovery had been announced the night before and it is somewhat surprising that the staff remembered to present the prize at all! Carl Lampland went down to award it.[16] At the same time, he announced the discovery of the new planet.

When Lampland returned to Lowell, he reported that the reaction to his announcement was muted. "It didn't make much of a dent," Tombaugh recalls. Moreover, the newspaper reported that it was Mrs. Lowell, who actually was in Boston at the time, who had made the presentation. Lampland was an indifferent speaker with a quiet voice; it is possible that the audience didn't hear him or that his delivery didn't carry enough urgency to make the people grasp the importance of his words.[17]

The announcement was out in time for the Thursday evening newspapers on March 13. By the following morning Flagstaff's telegraph office was flooded with messages. "We are harried and rushed with things for the press—" Lampland wrote that day, "telephone calls, telegrams, etc. Then I went to 42-inch and photographed X."[18] With minor variations from paper to paper, the Associated Press story ran with a mixture of romance and fact, as in the version that appeared in the Allentown, Pennsylvania, *Morning Call*:

Flagstaff, Ariz., March 13. In the little cluster of orbs which scampers across the sidereal abyss under the name of the solar system there are, be it known, nine instead of a mere eight worlds. The presence of a ninth marcher in the retinue of the Sun, long suspected, was definitely announced here today by Dr. V. M. Slipher of the Lowell Observatory, who headed a group of eminent astronomers whose gropings in the milky way with telescopes and cameras located the new sphere. Way out behind Neptune, tagging bashfully behind his brothers, this new planet's exact whereabouts, size and age are still unknown, and it hasn't even got a name.[19]

At first Tombaugh's role in the discovery did not get much attention. The Associated Press credited the discovery to V. M. Slipher, C. O. Lampland, E. C. Slipher, J. C. Duncan, K. P. Williams, E. A. Edwards, and T. B. Gill. John C. Duncan was involved in the first Planet X search in 1905 and 1906; Kenneth P. Williams searched in 1907; Earl A. Edwards and Thomas B. Gill helped first in Lowell's Boston office computations and later at the telescope, as part of a later search.[20] Not until the eighth paragraph does the story say, "First notice of the body was made by C. W. Tombaugh, photographer of the Observatory, who saw a tiny spot on one of his plates. Astronomers soon declared it to be the long-sought planet."

Near Burdett, Kansas, the weather on March 14 was fairly mild. Roy Tombaugh, a senior in high school, was eating lunch in his car with some friends. Suddenly, a friend sped down the dirt road, stopping beside him, his two rear wheels braking hard and sending up a cloud of dust.

"Why didn't you tell me about your brother?"

"What about my brother?"

"Why, he discovered a new planet! The newspaper people are going to the farm, and it's all over everywhere."[21]

The news disrupted the school, the town, and the state, for a Kansan had discovered the new planet. Charles Dilley, school principal, immediately called an assembly of all sixty students of the various grades. Trying to motivate his students, he had challenged them only a few months earlier: "What has this town ever done? Have there been outstanding people in this town?" He must have been deeply thrilled to retract these words. Later he worked to get Tombaugh a scholarship to the University of Kansas. "He poured himself out to the students that afternoon," Roy recalls. "I sat there. I was half numb. I couldn't really realize what was happening. I couldn't grasp the significance of it."[22]

Dinner conversation at the Tombaughs' that evening was a lesson in the solar system and its latest addition. Leslie Wallace, the publisher of the Pawnee County newspaper, *The Tiller and Toiler*, began a series of visits to the Tombaugh farm that day to gather some background for stories. He became a useful liaison between the Tombaughs and the press and helped them handle the pressure.

Of all the nation's newspapers, *The Kansas City Star* made one of the most intensive attempts to cover the story of the discovery of Pluto. One of the many people who looked at Tombaugh's picture in that paper was a high-school student named Patricia Edson. "That's where I first saw Clyde's picture," Tombaugh's future wife later recalled. "I thought he was kind of a nice looking guy."

The Orbits

Included in the storm of congratulatory telegrams were some hail-stones: requests for precise positions for orbit calculation. The Sliphers had hoped to produce the first orbit themselves, but now that the world knew of the new object, pressure was mounting exponentially to learn its orbit.

Buying themselves some time for their task, the astronomers re-fused to allow early positions of the object to be released, although that seemed to be a weak defense since anyone could now observe Tombaugh's discovery. Some people did, and others went to unusual lengths to "steal" positions from Lowell Observatory photographs. Harlow Shapley reported that one of his students had seen a newsreel that featured Director Slipher explaining a March 1 plate. (Lampland noted that the "sound movie people" had spent all of March 15 at the observatory.)[23] By frantically copying the field stars with the position of the planet during the few seconds the photo was visible, the student obtained "the first newsreel position of an astronomical object."[24]

A similar event took place at Poland's Krakow Observatory, whose director, Tadeusz Banachiewicz, had developed the "Banachiewicz method" of orbit calculation and wished to determine the new object's orbit. However, he had only the March 12 position announced on the Lowell circular and one other. He needed a third, and he found one in a photograph published in a newspaper shortly after the discovery![25] At the time, positions for astrometry were measured only from glass plates, and even today there is some debate about the accuracy of posi-tions measured from a flexible film surface. Measuring a position from a photograph published in a newspaper is dangerous by any standard, but that was all he had. The orbit he derived was almost circular, "wild," as Tombaugh would say, but not as inaccurate as some of the others!

These two stories illustrate the extent to which astrometrists went to obtain accurate positions for the new object's orbit computation. Humorous at first glance, they also demonstrate a surprisingly provin-cial attitude at Lowell. The scene of Lowell's director being filmed with a plate for the public while withholding the precise positions of the planet on that plate from the astronomical community is inappro-priate, but events like that happen today. In 1989 news of some new satellites of Neptune discovered by Voyager II was released to the press before being transmitted to the International Astronomical Union's Central Bureau for Astronomical Telegrams.[26]

Waiting twenty-three days to announce the discovery was a wise strategy that gave the Lowell astronomers a chance to confirm the nature of the object. Once the announcement was made, however, the observatory should not then have withheld precise positions, even though Cogshall (who had lent them the small camera attached to the 13-inch) suggested that neither Harvard nor anyone else had the right to insist on precise positions.[27] Cogshall was technically correct. Nonetheless, scientific etiquette demands a greater level of cooperation with the community. V. M. Slipher's circular of March 13 claimed that "it has appeared a clear duty to science to make [the object's] existence known in time to permit other astronomers to observe it while in favorable position." The observatory should have passed on its positions to Shapley at Harvard for immediate publication in the *Harvard Announcement Cards*; that was manifestly part of its "clear duty to science." Holding on to the positions was the Lowell astronomers' main weapon in a battle to produce the first orbit, a race they ultimately lost. Unfortunately, their policy also brought the observatory needless ridicule.

In any case, all the early orbits were inaccurate, since the arc of the observations of the new object was so short. "All the work that went into that first orbit," Tombaugh says, "was a waste of time." Not really: the preliminary orbit was good enough to permit a search for images on earlier plates.

In letters to Putnam, Shapley, and others, V. M. Slipher tried to justify his policy, essentially on the grounds that the discovery of the planet was the climax of a quarter-century's search and that the observatory was entitled to the chance to calculate the first orbit. Other astronomers were suggesting that the planet was not Lowell's Planet X and that perhaps it was not a planet at all but an asteroid or comet. Armin Leuschner, director of the Students' Observatory of the University of California at Berkeley, even questioned publicly in a San Francisco newspaper how the Lowell scientists could possibly know, without an orbit, whether this object was trans-Neptunian or not.

Actually, except for the trans-Neptunian nature of the body, which Tombaugh had established from its rate of shift, questions about the object's nature were appropriate scientific inquiries. Unfortunately, a hungry press gave these queries large headlines at the expense of Lowell's reputation. "Slipher was madder than hell at some of these guys," Tombaugh remembers, "and I don't blame him. I had a chance to see this pack of wolves, and I was appalled!" V. M. Slipher admitted that he was not about to let anyone overshadow the observatory's hour of triumph. The reluctance to provide precise positions was a human, not a scientific one.

One of the first requests for positions was from Leuschner, who was so anxious to get working on an orbit that he wired Lowell Observatory quickly after the announcement, asking for precise positions. Leuschner might have fared better with Lowell had he bothered to include a personal note of congratulations, but none was offered. Outraged by Leuschner's attitude, the Lowell staff did not respond. Leuschner may have realized his mistake, too, for only a week later he repeated his request for positions, this time in a letter filled with compliments about the achievement. It still did not mollify Slipher, who was annoyed with Leuschner's postulations about the new object's cometary or asteroidal nature. Nonetheless, Leuschner was an excellent celestial mechanician, and with two graduate students, Ernest Bower and Fred Whipple, was all set to compute an orbit.

When they finally received positions, they started work. The arc they had, defined by the first and last positions, was several weeks shorter than the two months used by Lowell. "We eventually produced several orbits," Fred Whipple said later, "and some of the elements were not far from the final values; not bad for such a small arc!" [28]

The senior staff at Lowell had not done much orbit computation since their student days at Indiana University. An answer to this problem appeared to lie with V. M. Slipher's former professor, John Miller, who was now director of Sproul Observatory at Swarthmore College. After some hesitation he agreed to assist at Lowell for a short period. Tombaugh remembers Miller, the Sliphers, and Lampland working virtually around the clock using tables of six-place logarithms. When the elements were finally computed after a furious four-day period, the resulting preliminary orbit also turned out to be wild—almost a parabola, with a period near three thousand years! E. C. made a three-dimensional model of it, and Tombaugh recalls, "It was a horrendous looking thing. What they got was so unexpected that they wondered how Lowell's planet could be that kind of a critter! Of course, Miller was wrong about that; he should have realized you can't calculate a reliable orbit on so short an arc." It was unfortunate that they still had not found the object's image on the 1929 plates, for a very good first orbit could have been computed from the longer arc that the earlier image would have produced.

The consternation about the orbit caused Tombaugh to worry about his own role in producing the observations. Had he recorded the time he started the exposures correctly? Any significant error in time would mean that the object was in a different position from the one they assumed it had at the time of the exposure and would make the resulting orbit calculation incorrect. "I worried that maybe some of it was my fault."

Lampland had confided another problem to Tombaugh: "the star [and planet] images have small cores that are slightly eccentric to the rest of the image. 'What do I measure?'" For brighter stars, he would not see these off-centered cores, so that the brighter stars would be systematically off from the fainter ones. This small discrepancy would result in a very wrong orbit. Tombaugh's fear did not last long, for most of the orbits produced by Leuschner, Whipple, and Bower, using different images, were also wild. Tombaugh was puzzled over those results, but "that was the nature of the beast, and when better ones came out, everything changed. And only then did I realize that it was not that kind of eccentric rascal. I felt a lot better about it then." The orbit controversy was straining everyone at Lowell; they felt almost under siege. Finally, in a *Lowell Observatory Observation Circular*, further precise positions and orbital elements were released on May 1.[29]

These early orbits were not true representations of the path of the planet all the way around the Sun, but they were accurate enough effectively to predict other positions within a year or so of the observations that were already known. That was the real advantage of these early orbits; they enabled others to search for prediscovery observations—that is, images taken on earlier plates but not discovered when these plates were initially examined. The first ones were finally found on the Delta Geminorum plates that Tombaugh had taken in 1929 and that the Sliphers had blinked, missing the tiny planet. "That kind of burned them up," Tombaugh says, "but of course they were looking for something ten times brighter and they went a little too fast." In any event, since the object was nearly stationary on those plates, finding it was very difficult.

An important early position was revealed by A. C. Crommelin, a highly respected celestial mechanician, on a plate taken on January 27, 1927, at Belgium's Uccle Observatory,[30] which extended the arc enough that a much better orbit could be determined. When the results of Crommelin's work were announced, the orbit was clearly more "planetary" than "cometary," possibly not belonging to Lowell's Planet X, but certainly sufficient to establish the status of Clyde Tombaugh's object as a planet. At the end of June the object was found on two old Lowell plates, taken on March 19 and April 7, 1915 — a significant find since it would extend the arc of orbit even farther back. The orbit based on positions that included these images refined, but did not substantially change, the orbit Crommelin had calculated. Finding these images was a sad event in a historical sense, for Percival Lowell never knew that his own search had succeeded in recording the new planet in his lifetime.

It was a special pleasure for Tombaugh when V. M. Slipher showed him a letter from Crommelin congratulating the observatory and specifically mentioning Tombaugh's role. Feeling honored that Crommelin had realized the nature of the search, Tombaugh recalled that no other astronomer had bothered to congratulate him at the time. During the year after the discovery, searches for new positions were being made both through the plate vaults (for old positions) and through the telescopes of many observatories. The extensive archives of Harvard Observatory contain a large series of wide and narrow field plates taken during the 1930–31 observing season and centered on the new planet.

A prominent celestial mechanician, Ernest W. Brown of Yale, was quite skeptical about the relation of the new object to Percival Lowell's Planet X. Convinced that the Lowell Observatory had located the new planet by accident, Brown claimed that its close resemblance to prediction was a happy coincidence. It was not until the discovery of Pluto's moon Charon by James Christy in 1978 that Brown's unpopular statement was finally shown to be correct.

Of the two good plates of January 23 and 29, 1930, only the January 29 exposure remains in the Lowell Observatory plate vault. The January 23 plate, by request and after some negotiation, now is at the

Smithsonian Institution. The Smithsonian had asked Director Arthur Hoag for both plates, but on Tombaugh's recommendation he sent only one. Had both good plates left, the observatory would have a break in the sequence of sky photographed with the 13-inch astrograph.

Minerva, Cronus, and Pluto

If the nature of the orbit was at the heart of the scientific controversy, the naming of the planet dominated the public one. After many names were suggested, the three final choices were Minerva, the Roman goddess of wisdom; Cronus, the son of Uranus and father of Neptune; and Pluto, the god of the lower world. Minerva was the favored choice until it was realized that asteroid 93 was already named Minerva. Cronus was briefly considered, but it was the suggested name for a planet predicted by T. J. J. See, a highly eccentric and egocentric astronomer who, it was feared, would claim the planet as his if that name were used.

An eleven-year-old named Venetia Burney, from Oxford, England, seems to have been the first outside the observatory staff to have suggested Pluto, and Slipher did credit her with the name in his announcement of the name. Tombaugh, remembers, however, that though this gifted suggestion was used, the staff had come up with the same proposal about two weeks before learning of it. At the time the only objection to the name Pluto, alluded to by Putnam, was its use in Pluto Water, a popular mineral water laxative.[31] The relation to Walt Disney's floppy-eared dog apparently was never mentioned, and Tombaugh does not recollect any conversation about it from that time, adding that the planet could have been so named "because it was so doggone hard to find!"

In fact, Pluto the dog did make his first brief appearance in 1930, in a Mickey Mouse cartoon called "The Chain Gang," and his own cartoon series was started in 1937.[32] Finally, on May 1, Slipher proposed the name Pluto as well as the symbol, a stylized *P* and *L*, an inspired notation that is far easier to write than the complex symbols for Uranus and Neptune and, at the same time, honors the memory of Percival Lowell through his initials (see page 9).

A sidelight to the naming fray involves Percival Lowell's widow, who had suggested Zeus for the new planet, an idea she quickly amended to Percival. That name lasted but one day in the mind of Constance Lowell, who finally proposed Constance on a somewhat whimsical note: "Are you willing to have the planet named Constance?" [33] Whether Mrs. Lowell was serious or not is unknown; Tombaugh was "furious" at the suggestion, and one might imagine that the rest of the observatory staff did not think very highly of it.

Back to Work

With all the excitement, it was difficult to think about getting back to work that spring. Requests for articles about Tombaugh's life began on March 14; Lampland noted that Tombaugh "was writing an article for Science Service—sketch of his life, I think." [34] Many astronomers visited Lowell, and Tombaugh showed them the two discovery plates, which were still on the comparator. One astronomer was amazed that Tombaugh could have picked out the interloping planet; he could barely see it, even though it was clearly marked. Tombaugh answered that he habitually checked things several times fainter. Roger Putnam finally was able to visit, and Tombaugh enjoyed his company immensely.

Near the end of April, Milton Humason visited Lowell while conducting seeing tests for the planned 200-inch telescope at Mt. Wilson. One of Mt. Wilson's best-known astronomers, he had started in that observatory's early days as a mule driver and progressed through observatory janitor to night assistant, to the full status of astronomer. He reported the results of a two-hour spectrum exposure of the new planet through the 100-inch Hooker reflector. Pluto had a solar-type spectrum, indicating a planet, but unlike the giants Uranus and Neptune, there was no evidence of an atmosphere. Since Humason had done some searching in 1919 for William Henry Pickering's predicted Planet O, he showed considerable interest in Tombaugh's set-up and was particularly amazed at the large numbers of stars Tombaugh had to search through.

The spring of 1930 left Tombaugh quite busy, but without a specific task. Although he was not involved in the orbit calculations, he did help wherever possible. V. M. Slipher kept him up to date on what was happening and on what other people were saying. Tombaugh could not do any further blinking, since Lampland had inserted some 4-by-5 inch plates from the 42-inch and was blinking these planet images.

By the middle of May the idea of continuing the search was gaining strength. Tombaugh was still taking plates, and he began to think that since the first discovery had come quickly enough, maybe there would be more. Continuing the search would find them if they were there; he expected to find two or three other planets with the powerful 13-inch.

If the uncertainties over Pluto were not enough, on April 22 Canada's Dominion Observatory reported a second trans-Neptunian planet; apparently, it had been found on plates taken six years earlier in a search for early images of Pluto.[35] If the object were real, more positions would be needed to establish any trans-Neptunian status. As Tombaugh resumed blinking the Delta Geminorum plates, he remained especially watchful for what had by that time become known as the "Ottawa object" and found no evidence of it. "It irritated Lowell [Observatory] a bit," Tombaugh recollects. "They were getting split attention on planets!"

With interest focusing on possible additional planets, V. M. approached Tombaugh late in May to request, not to Tombaugh's surprise, that the search continue. The procedure would be as before; Tombaugh would take the plates and blink them. So Tombaugh removed Lampland's micrometer eyepiece, inserted again his standard blinking eyepiece, found the Pluto field, matched the plates carefully, turned on the motor and adjusted it to three blinks per second, and then picked up where he had left off three months earlier.

As Tombaugh became accepted as an equal member of the staff, the senior members told him stories about the observatory and its founder. When the planet search resumed, Tombaugh was the only one involved in it. Despite the years of acquaintance, however, Carl Lampland still addressed Clyde as Mr. Tombaugh, and E. C. Slipher used the incorrect pronunciation "Mr. Tombo." Alone among the senior staff, V. M. Slipher called him Clyde. By this time the Lowell staff had the confidence and the good sense to let the entire program be the domain of the young Kansas man who would very likely never be a farmer again.

Summer of 1930

With the coming of the summer rainy season, V. M. Slipher let Tombaugh return home. The pressure on the observatory was beginning to slacken, with one important exception: many times more people than usual were including it as part of their summer vacation trips. Always an ideal summer spot, in 1930 Lowell had the special attraction of being the place where Pluto was discovered. In the meantime, Tombaugh was invited, in August 1930, for lunch and a meeting with Arizona's governor.

Tombaugh recalls a tense week-long visit by Constance Lowell that summer. With the extra pressure on the observatory, this visit was especially difficult. Still dressed in mourning black, she had a commanding presence that unnerved the young Clyde. Mrs. Lowell had told V. M. Slipher that she was anxious "to meet the young man who had found my husband's planet." V. M. walked with her to the comparator room and left her alone with Tombaugh. No one had bothered to warn him about Mrs. Lowell's personality, although he was aware of the damage she had done to the observatory after her husband's death. The discussion began innocuously enough, as Mrs. Lowell asked Tombaugh what the planet had looked like. Then she began berating the director, accusing him of having cheated her "out of the discovery of Planet X." Tombaugh believes that she wanted him to support her position, perhaps even to link his discovery to whatever she may or may not have seen. Trying to protect himself in an impossible situation, he said as little as possible. "It put me into an extremely awkward and tense situation, and I just thoroughly detested her for that."

For three weeks in July, Lampland, with some assistance from the Sliphers, handled the crush of visitors in order to allow Tombaugh a visit home. He looked forward to threshing in July, a different kind of hard work that would clear his mind and bring him back to his roots.

Tombaugh was now so popular in Kansas that letters addressed to "Clyde Tombaugh, Kansas," had no trouble reaching him. Nevertheless, his train was not met by a marching band, just his family. Aside from a short visit to some relatives in Nebraska, Tombaugh's three weeks at home were devoted first to his family, second to the press. One afternoon the well-known reporter A. B. McDonald, of the *Kansas*

City Star, visited the farm with Leslie Wallace, of *The Tiller and Toiler* of Larned, Kansas. McDonald had already published an article, complete with Tombaugh family photograph, and this trip was a return visit to look at the telescope. Clyde and his brother Roy were working in a field when Roy noticed the newsmen and suggested that he and Clyde go back to the farmhouse. Roy recalls: " 'Oh, yeah, let them wait,' Clyde told me; he wasn't going to quit cutting wheat!" [36]

Chapter 6

THE THIRTIES

B ASKING IN the success of its planet search, the Lowell Observatory had leaped into its golden era in a time that happened to coincide with one of the most difficult periods of this century. It began just before the October 1929 stock market crash, reached its peak of intensity during the depression's worst years, and ended with the Second World War.

The depression affected the observatory directly in that its budget, still strained from Mrs. Lowell's legal antics, continued to be very tight. Not helping matters was an incident involving a local bank whose president vanished with about a million dollars of funds. The observatory lost money in that debacle, and Tombaugh himself lost about one hundred dollars, almost his full monthly salary at the time. The Great Depression was worsening, and people Tombaugh knew were unemployed. Compared to that, his salary plus room at Lowell was wonderful.

For Clyde Tombaugh in the summer of 1930, world affairs were clearly far in the background of his own personal achievement. Life was full, each day busy with blinking and public tours, and many nights taken up with observing. Wilbur A. Cogshall visited from Indiana University and was teaching a university student named Henry Giclas to use the 13-inch. The next year Giclas was hired to assist with various duties around the observatory. With the work now so much more in the public eye, and with Tombaugh himself very busy with the planet search, it was thought prudent to hire someone new. Having someone of his own generation at the observatory was a new pleasure for Tombaugh, and soon Giclas was taking occasional 13-inch plates.

By this time Lowell was certainly getting its money's worth out of Tombaugh. Once the decision was reached to continue the search, the program greatly expanded. First, the idea was to continue the search around the entire zodiac region, and quickly the idea developed to double the work. To speed up the search, Tombaugh decided to search on both sides of the zodiac strip he had completed the year previously. By the start of the summer storms in 1931, he would thus have completed a strip around the zodiac that was thirty-five degrees wide!

Observing Notes

A successful observer has to take nature by the tail. When clouds interfere with observing, a resourceful person will have everything ready to go and be watching the sky so that observing can begin immediately when it clears. Tombaugh spent many hours waiting, reading, and periodically opening the administration building's door to check on the condition of the sky. The 13-inch telescope dome would be closed, but a plate was loaded in its holder and ready for an exposure in case the sky cleared.

A region more than fifteen degrees from the opposition point was too far for effective blinking. Since the opposition point moved about twice that far each lunation, the schedule was demanding; for best results, at least two of the three plates of each region needed to be taken in the same week, and about thirty one-hour exposures were required in each lunation. Keeping up with this opposition point is even more difficult when the plan involves taking three hour-long exposures of each region. Further, Tombaugh tried to keep two nights between plates of each region, which made the schedule even tighter. Because of uncertain weather, in a few sets the interval between plates was only one day. A run of bad nights forced some regions to be dropped from the program, to be resumed the following year. One advantage of taking three plates of every region is that it really doesn't matter which two are blinked, leaving the third for a check plate. This way a problem with clouds or poor seeing might prevent a plate from being blinked against another plate but still render it acceptable as a check plate.

Occasionally, at the end of one lunation Tombaugh would have only two of his required three plates for one region. With the waxing gibbous Moon encroaching on his dark sky, he would abandon the idea of getting a third plate until the following lunation. This plan was workable but not ideal. In some cases there would be a few days' difference between two of the images, whereas the check plate taken the following lunation could be three weeks away, with its suspect object getting farther from opposition. "The reason I preferred the shorter interval is that I didn't have to look as far to find those fainter images."

A total eclipse of the Moon once provided a chance of dark sky for a critical third plate. Tombaugh knew that it would be a long eclipse, but no one had any idea how dark the Moon would be during the total phase—bright orange, shades of light to dull red, or very dark. It is the dust content of Earth's atmosphere that determines this brightness, and after a major volcanic eruption on Earth, the Moon can be almost invisible, leaving a very dark sky during totality. This eclipse turned out quite dark, with the Moon's total light reduced to that of a third magnitude star. The sky had become dark enough that, for over an hour, Tombaugh took the plate he needed.

At that time Tombaugh's interest in eclipses must have been far less than his desire to keep to his plate schedule. Since the Moon was near the area he was photographing—at full phase the Moon is always in the opposition region—he could see the progress of the eclipse through the dome's open slit. "I looked at it, and it was dark and all that, but I kept my eye close to the guiding eyepiece. It was pretty dark, so I got a good plate." Later someone teased Tombaugh about his plate, calling it "a heck of a way to abuse an eclipse!" Tombaugh agreed: "Wasn't that an awful way to treat an eclipse of the Moon?" With the Moon emerging from total eclipse, the sky brightened and Tombaugh ended his observing session, feeling good that he had somehow taken advantage of a loophole of nature.

A slightly hazy sky required a longer exposure in order to bring up the magnitude limit to match that of a plate taken under more normal conditions. Prolonging an exposure too much had another deleterious effect. If the stars in the longer exposure were far enough away from the meridian, a refraction effect resulted, in which the thickness of atmosphere near the horizon acted as a lens, bending light rays from stars and changing their apparent positions. The amount of refraction depended on a star's altitude above the horizon. Differential refraction added to the difficulty; stars of different colors were displaced by differ-

ent amounts. These effects made matching star images from one plate
to another impossible.

A passing cloud presented a different problem. Tombaugh did not
want to close the exposure temporarily, reopening the shutter after the
cloud passed by, because accurate start and end times for each exposure
were critical if precise positions needed to be measured for any newly
found object. Usually, plates taken under problem conditions ended up
only as check plates. If the plates were not perfectly matched, Tombaugh maintained, blinking would become such a mental aggravation
that "in five minutes I would be practically screaming!"

The routine became even heavier each June as the summer rainy
season approached. During the storm season of July and August, when
moist air flowing from the southeast dominated the weather pattern,
one could not count on any nights for observing, although in some
years a stronger westerly air flow allowed some work to be done. During the last nights of June, Tombaugh would go as far east of opposition
point as he could in an effort to get a jump on the weather so he would
not lose too much sky around the opposition point. Often those nights
would feature intermittent high cloud, the remains of distant thunder-
storms to the southeast. Under these conditions Tombaugh couldn't
rely on an uninterrupted hour of clear sky, but he couldn't afford not
to try, just in case the sky did stay clear throughout an exposure. Tombaugh remembers these nights as being much more difficult than clear
nights. The plates had to be loaded, and then he would wait as clouds
passed by, checking the sky every fifteen minutes or so. As many observers realize, it is better to have either totally clear nights or totally
cloudy nights. The lazily scattered clouds of late June were frustrating.

Guiding a photograph with the well-mounted 13-inch was easy work,
but the hour-long exposures were tedious. With no radio until the
fall of 1936, he would avoid boredom by doing mental exercises of
the type he had enjoyed as a farmer. He would develop telescope systems, including calculating mirror sizes and focal lengths. Sometimes
he might lose his train of thought, so he would recalculate, all the while
leisurely guiding a sixty-minute exposure. "We had had no music at
home, except my mother played the piano. So I lived in a silent world;
I was all for seeing and looking." Once the radio arrived, late-night
accompaniment was mostly from Spanish-speaking stations.

The longest exposure he took lasted five hours, to capture the galaxies in the densest part of the Virgo region. When he had processed this plate, he walked with it down the hall to V. M. Slipher's office. It was probably the most beautiful plate of the Virgo region that the director had seen; Slipher was "delighted; his mouth just watered." Actually, continuing the exposure beyond three and a half or four hours bought little more data, since he had reached the level of the brightness of the background sky. In the final hour all he had earned was increased fogging from the background of sky.

The survey eventually went as far south as minus fifty degrees. A problem with going so low in the south was that some tall ponderosa treetops physically blocked off the view below declination thirty-five degrees south. Early in the spring of 1937 Tombaugh felt that covering the southernmost area of the zodiac in Scorpius, Sagittarius, and Capricornus would be possible if the tops could be trimmed off some trees. It would extend the coverage southward even in areas far from the ecliptic by perhaps fifteen degrees or thirty Moon diameters! One afternoon Tombaugh took V. M. Slipher on a tour during which he pointed the telescope to fifty degrees south and moved it an hour east, then an hour west, of the meridian. He then marked the trees that needed trimming. A few days later a work crew came by and cut almost twenty tons of treetops. Tombaugh succeeded in taking the southerly plates in June.

Taking these far southern regions when they were on the meridian, as high as they would ever get in the sky over Flagstaff, wasn't quite sufficient; their stars were so low that they suffered from refraction as well as absorption by large amounts of atmosphere. Tombaugh found that for these regions the 5-inch Cogshall camera plates were much better to use for blinking. Although they had only a third of the scale, this factor was now an advantage because the refraction would be only a third as much. Thus, except for the rich Scorpius and Sagittarius regions, the regions from minus forty degrees to minus fifty degrees were photographed and blinked using the Cogshall instrument.

Tombaugh defined his search as involving the entire sky visible from Flagstaff, except for angular distances greater than sixty degrees from the ecliptic, for any object at Saturn's distance or greater. Since an object as close as Saturn would show considerable motion from one year to the next, in order not to miss it Tombaugh had to overlap his exposures so that it would not get lost over one year. The amount of overlap varied from up to 50 percent in a north or south direction, to somewhat less in east or west.

The pressure of keeping to a schedule involving all these conditions was overwhelming. Rarely did Tombaugh get so exhausted that he would doze off during an exposure, although his notes record that it did occasionally happen—"but only for a minute," Tombaugh insists. Despite the fatigue, only once did he confuse plate holders and expose different regions on the same plate. After a long winter's night in the frigid dome and sometimes only four or five hours of sleep, the work would be faced again the next night. "I remember," Tombaugh adds, "if I had clear weather, as the Moon approached first quarter it was a welcome sight. I was so thoroughly exhausted I did nothing but lie around for a few days." It would be a while before he could even use the blink comparator, and he would sometimes not even want to read.

Notes on Blinking

Switching from the nighttime observing routine to daytime blinking was difficult. Because blinking at night, with room lights on, produced a glare that Tombaugh found tiring on his eyes, he needed to blink in a room lit by daylight. Moreover, "when you spend a lot of time in a cold dark dome, you have to see some light for a change, some warmth, so I would rather have blinked in the daytime. You get tired of this dark silent thing at night."

It has been suggested that the area of sky actually blocked out by star images covers 2 percent of the sky and that there is hence a one in fifty chance that Tombaugh might have missed something because it had been obscured by the light of a star. In response, Tombaugh assured the author that if an object appearing on only one of two plates was near any star, he would invariably bring out the third plate to see if there were an image on the star's other side. Emphasizing that he always considered the possibility that his prey was lurking in front of a star, he checked every such case.

Immediately after a plate was processed, he would usually scan it quickly for about ten minutes "with a hand magnifier to see if anything unusual, like a bright comet that required immediate attention," was there. He would then file the plates into a queue for blinking, a process that might not take place for a year or more. In 1934 there was a water shortage, which gave Tombaugh some concern that the plates were not being washed thoroughly enough. After the shortage was over, he spent many hours rewashing all those plates, recording on their envelopes when and for how long they were rewashed.

During blinking, any change from one plate to another, other than an obvious plate flaw, would be marked. The fainter images revealed on the 2.5-hour exposures that were taken late in the survey showed many doubtful objects that could not be separated from plate flaws. Each one was checked out as a suspect. On the back of each plate Tombaugh would outline each suspect with a pair of green ink dots and an appropriate letter to identify the probable nature of the object. He marked *A* for asteroid, *V* for variable star, *T* for temporary object or nova suspect, *N* for nebula, *DN* for double nebula, and *RN* for remarkable nebula. On one of the 2.5-hour plates he marked ninety-nine asteroid images.

Other than recording the asteroid images, Tombaugh had little time to check them out to see if they represented new ones or previously discovered ones. By 1931 Carl Lampland had begun a system in which assistants Alan Cree, Henry Giclas, Kenneth Newman, and Margaret Purcell would record the positions of all the asteroids against an overlay chart. Approximate positions of hundreds of asteroids were obtained in this manner and were then compared with orbital elements of known asteroids to see which were known and which were not.

Years later Fred Whipple, now one of the world's most highly respected comet scientists, suggested that there could be a small chance that an Earth-approaching asteroid, or some other object on a strange orbit, might appear at a certain distance from Earth to show a displacement similar to that of a trans-Neptunian planet. Tombaugh comments, "I told him I did not find any. Of thousands of suspects, not a single one turned out to be such an object. But I don't think I convinced him."

As the years of examination progressed, Tombaugh's respect increased for the many types of objects the sky had offered him. Realizing that he had a unique way of observing the structure of the galaxy, he would record the galactic latitude and longitude (a coordinate system based on the galactic equator or plane—the middle line of the Milky Way, which is inclined sixty-two degrees to the ecliptic, an extension of the Earth's orbit into the celestial sphere). He wanted a picture of what each region offered in its relationship to the plane of the galaxy.

After each set of plates was blinked, on the jacket Tombaugh recorded extensive notes that described every find, every suspect, every unusual thing that appeared. He would write these notes immediately after blinking was completed. The notes included projections of star counts, done by placing a 14-by-17-inch mask to cover the entire plate. The mask was pierced with 35-by-10-millimeter holes spread evenly across its surface. He would then count the stars within each hole. The average count, when multiplied, would come to within 5 percent of the total number of stars on the plate. Using this procedure, Tombaugh calculated that each of two plates, one in Sagittarius at the center of the galaxy, the other in the Scutum star cloud, contained more than one million star images! On the plate envelope of a rich region centered on 27 Scorpii, Tombaugh wrote, "10 days of hard, solid work." On a typical day he might blink thirty to sixty thousand stars.[1] During his entire survey he blinked about ninety million stars "plus or minus one million."

In September 1936 Tombaugh recorded a comet suspect that he positively identified on the slightly better September 22 plate to be two asteroids, A1 and A2. A2, which looked a bit defocused, was still a comet suspect. "The trail is fairly long, very faint in intensity and somewhat fuzzy which gave suspicion that it might be a comet. Finally after 3 hrs. work chasing around I gave up."[2]

Tombaugh did discover two comets, but neither was reported, since they were on plates taken more than a year earlier (see appendix). The only comet discovered and actually announced during the search was in 1932, when Kenneth Newman found a comet in Ophiuchus on a plate exposed June 20 and confirmed on two other plates taken June 1 and June 7. It is known as Comet Newman 1932 VII.[3]

Planet Suspects

The meticulous care Tombaugh took in his program is amply demonstrated by his handling of suspects. Early in June 1932 he found a "promising planet suspect, 16½ magnitude," on plates 410 and 412, a pair taken more than a year earlier. However, the "object" was very close to the plate limit. To check this suspect, Tombaugh took a ninety-minute exposure on June 6. The new plate showed a "definite star image" where the possible planet's image had been before. Tombaugh ended the investigation with a philosophical note on the plate jacket: "There the suspect fails and the matter is dismissed."[4]

On July 4, 1932, Tombaugh encountered "the most intriguing planet suspect since Pluto." It appeared to exist, along with 130 other "questionable objects," on plates 419 (taken May 12, 1931) and 421, centered on 27 Scorpii and "confirmed almost perfectly in position angle and shift on the 3rd plate." He reexposed the region in early August 1932 and still couldn't distinguish between real images and spurious ones. Finally, he remeasured the shifts on the original three plates. The result was that the one-day shift from May 12 to May 13 was "14 ⅓ twentieths of a mil[limeter]," but that the daily shift between May 9 and May 13 was "¹⁷⁄₂₀ths of a mil," which placed the May 9th image "¹¹⁄₂₀ths of a mil too far east."[5] Quite possibly, a less careful investigator would have reported this image as a new planet at about Neptune's distance.

A third promising object was found around August 24, 1933, on plates 484 and 487, centered on the star 37 Aquarii. Its shift was about the same as that of an object at the distance of Uranus. After much work Tombaugh ended his notes on the suspect thus:

> This promising planet suspect has met its fate also. Thinking that its brightness might be great enough to show on the 5-inch Cogshall plates, they were consulted. The image on one plate was confirmed, but not the other one. The 13-inch plates of 1929 were then put on the comparator with plate no. 484, and there were very faint images in both positions on both plates. Then it was recollected that this suspect would be bright enough to show on the weaker Eastman Speedway plates of an overlapping adjacent region to the N.W. Putting these plates on the comparator with plate 484, there appeared in the exact positions of the planet suspect images, stars in both positions on both plates strong enough to make certain that these planet suspect images were only real star-images, and the one star is probably a variable. *This is final.*[6]

Some planet suspects turned out to be RR Lyrae variables, stars whose brightness changes within a few hours. Two such stars out of phase with one another could together masquerade as a planet suspect.

These images had been taken on coarse-grained Imperial brand plates, which, although fast, tended to have too much "dirt" — actually random clumps of grains that would darken emulsions just enough to be mistaken for planet suspects. To increase the visibility of yellowish planets like Pluto, the Lowell staff members had thought for a time of using a large yellow filter, but the problems of keeping such a huge piece of unprotected glass clean discouraged them from trying one.

Clusters and Telescopes

The blinking session conducted on June 1, 1932, produced the second major discovery of the search program. Tombaugh was blinking a set of plates he had taken in May 1931. Plate 417, centered on Pi 68 Hydrae on May 12, was in a far southern region in the constellation of Hydra just north of Centaurus. He noticed what first appeared as a ninth magnitude star whose edges were less amorphous than other stars of similar magnitude. The "star" was in fact listed as a nonstellar object, having been discovered by Herschel in May 1784.[7] To Tombaugh the object looked like a globular star cluster. Thinking that it would be interesting to identify it, he checked Harlow Shapley's standard list of globular clusters and was surprised that this one was not included. He then checked a copy of the *New General Catalogue*, whose listing of more than seven thousand objects was completed in 1888 by John Dreyer.[8] This catalog did list the strange object as No. 5694 but described it as nebulous. The catalog almost made the connection to a globular cluster, but not quite, suggesting confusingly that it was "pretty suddenly brighter in the middle, resolvable (mottled; not resolved.)"[9]

As he had done more than two years earlier, once again Tombaugh walked across the hall to Lampland's office. "I think we may have a new globular cluster," he stated. After examining the image, Lampland decided to photograph the object with the 42-inch reflector. The large telescope confirmed Tombaugh's suspicion, and a cluster more than one hundred thousand light-years away was revealed. It is seen over our Galaxy's central hub, on the other side.

Announcing this discovery presented some problems, not the least of which was the cluster's integrated magnitude. With the 13-inch Tombaugh thought the magnitude was nine, but the value derived from the 42-inch telescope plate was closer to seven. Reducing the 42-inch plate to the same scale as that of the 13-inch did not bring a closer magnitude agreement. In the meantime, a Harvard plate taken with a small lens showed the object to be around eleventh magnitude. Finally, the announcement properly listed the different magnitudes as measured through the different instruments. The discovery of the ninety-fourth globular cluster was published by Lampland and Tombaugh in the August 1932 issue of the journal *Astronomische Nachrichten*.[10] A revised centennial edition of the *New General Catalogue* with 5694 correctly listed as a globular cluster appeared in 1988.[11]

The zodiac passes through the Milky Way regions twice, in Taurus and Gemini in winter and in the star-filled areas of Scorpius and Sagittarius in summer. It was in these star-filled regions that Tombaugh discovered five open clusters. Some were right in the center of the Milky Way, others at its edge. Each time he encountered what appeared to be an open cluster, he checked with Shapley's catalog to see if it was listed there and, if not, then referred to the *New General Catalogue* to see if it listed anything.

In 1933 Tombaugh built what appears to be the first "richest-field telescope" in the United States. This telescope had a very short focal length, enabling unusually wide-field views of the sky. In June, Lampland silvered the mirror for Tombaugh, and it became an instant sensation. Its magnificent views of the Milky Way provided many hours of thrilling views for the staff and the observatory's visitors. It is a 5-inch f/4 reflector that survives today as a finder, riding atop his 16-inch. "I would never make a telescope that short now," he advises, "not even an f/5. I would not want anything faster than f/6; otherwise, you simply get too much coma at the edges." This telescope is described and pictured in *Amateur Telescope Making, Advanced*.[12]

The following year Tombaugh made a 5-inch mirror of longer focal length. Since he was no longer using his first 8-inch from 1926, he simply cut the new mirror out of the old one with a coffee can. The other scopes that year were 12-inch mirrors of two different thicknesses. The thicker one turned out slightly overcorrected, but the thin one was almost perfect. "I was truly surprised at the performance of the thin 12-inch," Tombaugh recalls. He designed an intriguing support for it using eight ¾-inch strips of pine wood, all the same widths and exactly parallel to each other. Cardboard and a small piece of carpet sat on top of the wood, with the thin mirror resting comfortably on top of the carpet.

University of Kansas

When Clyde Tombaugh was awarded a four-year Edwin Emory Slosson scholarship to the University of Kansas, he accepted it as a fabulous surprise. Named in memory of a well-known chemist who had started his work in Kansas, the scholarship had just started, and the famous planet discoverer was the first recipient in the spring of 1931. He had not applied for it and had always despaired of ever having the financial means to attend a university.

The news was good for Lowell Observatory as well, for Tombaugh would return, the Sliphers felt, a more highly educated man and more valuable to the observatory. However, the scholarship was for the following autumn, a time Tombaugh had been looking forward to using to complete his intense double-strip photographic program of the regions around the zodiac. The feeling was that by this time other groups would be involved in planet searching and to slow the pace now would be foolish. On behalf of the observatory, he requested and was granted a year's delay in the start of his higher education. In the fall of 1932 Clyde Tombaugh, aged twenty-six, enrolled as a freshman student at the University of Kansas.

The astronomy program at the University of Kansas was run as part of the Physics Department by Dinsmore Alter, later a famous lunar expert, who had built a 27-inch reflector.[13] He had persuaded the university's Mechanical Engineering Department to design and construct the mounting. To help raise money for the telescope, he even had invested in the stock market, whose crash in 1929 ruined that plan. Had the telescope been successfully completed, Alter had intended to begin an asteroid survey, involving the locating and observing of known asteroids, obtaining precise positions, and calculating orbits.

Tombaugh saw introductory astronomy as a perfect "snap course" and was all set to register for it. He had no astronomy credits and thus was technically eligible to enter. Dr. Alter would have none of it. The idea of the discoverer of the solar system's ninth major planet sitting happily in the introductory astronomy lecture hall and answering questions about Kepler's laws seemed ludicrous to him. "I was disappointed because I thought it would be fun just to go through it," Tombaugh smiled. "But no such luck!"

Tombaugh's first year away from Lowell was an experiment for him as well as for the observatory. Although no one else would be permitted to blink, possibly others could at least take photographs while Tombaugh was in Kansas. A young assistant who had been hired earlier to perform different tasks around the observatory, including assisting Lampland with his 42-inch work, was given the job of taking planet-search plates in the spring of 1932. Before Tombaugh left, he supervised the selection of guide stars and the early photographic work. Afterward the other men, spoiled perhaps by Tombaugh's expertise and independence, left the new assistant essentially on his own. His first plate began the observatory's "X" series of plates, lasting from X-1 to X-180.

When Tombaugh returned that summer of 1933, V. M. Slipher and Carl Lampland greeted him with some consternation, asking him to take a look at the last year's plates as soon as possible. Although it took him about a day and a half to examine all the plates, it did not take nearly that long to discover that they had serious problems. The hapless observer realized that he had "made a mess of it," Tombaugh senses, for he came in frequently to ask about them. Typical of the kinds of errors he had made were that on two plates supposedly intended for blinking, one was centered on Capella, the other guided on a faint star more than five degrees away! "How could you make such a mistake?" Tombaugh inquired. "Don't you know the difference between a zero magnitude and a sixth magnitude star?" Also, some of the times, and even dates, had been incorrectly marked. By looking at asteroids whose orbits were well known and comparing their positions with the times recorded on the plates, he found that some could not possibly be where they were on the dates the young man had recorded. Only a quarter of the plates were suitable for blink comparator examination. Knowing how expensive the planet search was and how little money the observatory had to fund it, Tombaugh was embittered by this apparent waste of a year of observing. Reporting the news to Slipher was sad and difficult, and Slipher responded by firing the unfortunate observer.

Visiting Lowell that summer was a young graduate student named Frank Edmondson, who was working with V. M. Slipher on a mas-

ter's thesis on radial velocities of globular clusters as part of his Lawrence Fellowship. (This fellowship program had been set up by Percival
Lowell in 1905 to help students from Indiana University work at the
observatory.) Having no one to continue the planet-search photography, Slipher asked Edmondson if he would be interested in the photographic work at the telescope. Knowing the near disaster of the previous lost year, Edmondson approached this work with the seriousness
and thoroughness that characterized his later career as an astronomer.
"My responsibility begins here," Edmondson wrote on the opening
page of his part of the official plate log. He ended up taking plates
X-181 to X-480.[14]

Meanwhile, the blinking got further and further behind. Plates taken
in 1931, for example, were blinked in 1932, and by the time Tombaugh graduated, some of the plates were several years late being
blinked. Nonetheless, aside from the slight possibility that an observer
at some other institution might discover a planet ahead of the workers
at Lowell, Tombaugh saw no problem in the long delay between observing and blinking a plate. "That didn't bother me at all. I felt perfectly confident in retrieving anything I found."

Patricia Edson

Dormitory life for a studious freshman can be difficult. Although
Tombaugh did get along with the other students, playing touch football with them, for someone used to the discipline of farm work and the
quiet of observatory life, living in a dormitory with students seven or
eight years his junior was exasperating. He did enjoy the Syzygy Club,
a social club Alter had founded for his astronomy students. Named
for positions of the Moon when new or full, the group consisted of
some of the most talented of the astronomy students. In the spring of
his freshman year Tombaugh met James Edson, a bright and creative
astronomy student, through the Syzygy Club. Edson recalls that the
club provided a more serious outlet for students anxious to avoid the
"rah-rah" college atmosphere. The group had its own lightheartedness,
"an appetite for jokes, speculations about space travel, and hot dogs
cooked in front of the radiant gas heater in the astronomy classroom.

Tombaugh was a well-liked, rather quiet, and respected member of this small fellowship." [15]

When Tombaugh began looking for a place to live during the following academic season, James mentioned that his family had extra rooms to rent to students and others and invited Clyde to visit. James's sister, Patricia Irene, had been looking forward to meeting the young man who had discovered Pluto, but Tombaugh arrived rather unexpectedly. "I came down to the house, and Patsy was upstairs and she was sweeping and she had an apron on. She felt really embarrassed because she had her work clothes on, you see."

"I was a bashful girl twenty years old," Patsy adds.

Patsy's father had graduated in engineering from the University of Kansas, and he and his wife had hoped that each of their three children would get a university education. He died during the 1918 influenza epidemic. In 1932 the family moved to Lawrence, and Mrs. Edson got a job at the cafeteria so that the two boys could attend the university. "I was a girl, and it didn't matter whether I was educated or not," Patsy admits, "so I stayed at home."

Tombaugh liked the house and rented a room there. "He visited again in the summer," Patsy remembers, "just before he went to Flagstaff. That's when we really met. We went to the show a couple of times, and he kept coming around the house. Then he went to Flagstaff, and we wrote to each other all that summer." Thanks to a cousin's generosity, in the fall of 1933 Patsy finally enrolled at the University of Kansas, first in drawing and painting and, after two semesters, majoring in philosophy. She and Clyde took classes together in geology and astronomy, and they especially enjoyed philosophy.

Early in 1934 Clyde asked Patsy if she would like to accompany him to Flagstaff that summer. Patsy took that for a proposal. "I thought I wasn't going to go with him unless we were married; at least that wasn't done in our day! He never denied it was a proposal!" In June 1934, after spring semester, Clyde, twenty-eight, and Patsy, twenty-one, were married in her mother's home. Neither had many student friends, so there were more university professors than students as guests at the wedding.

Patsy's grandfather had felt that " 'Pat doesn't need to go to school; she should get married.' That stuck in my craw, you know," Patsy recalls. "When I got married and later graduated with honors, I wrote him a letter: 'How do you like that, grandfather? I did both.'"

Fall of 1934

As Patsy and Clyde were leaving Kansas for Flagstaff, they stopped at a filling station, and her husband was immediately recognized: "You discovered that planet!" It was a rainy, dreary day, and they were driving their small antiquated Ford, packed tightly with all their possessions so that there was just enough room for them to squeeze in. Patsy even had telescope parts on her lap.

"Did you get rich?"

They glared at the attendant. "Do you really think," Patsy answered directly, "we'd be driving across the country like this if we were rich?"

Tombaugh graduated from the University of Kansas in 1936 and then put in two further years of full-time work photographing and blinking at Lowell. During this time they saved money for a future attempt at graduate school, although it was difficult to save much. There was little money around at Lowell, and Tombaugh was loath to ask for any more of it. "I said to him," Patsy remembers, "it was a holy call like the ministry, and he was dedicating his life, soul and all, to science. . . . It was a pain in the neck." Tombaugh didn't disagree with that interpretation of his work, except to word it more as a sport than a science: "After all, I was hunting big game." In any event, graduate school would have to wait.

Patsy's first experiences with the somewhat rustic conditions at Lowell were unnerving but livable. After their wedding Tombaugh naturally moved out of his small room on the second floor of the administration building into a very small house on the Lowell grounds, with a long narrow front room, a tiny kitchen, and a bedroom so small that, Patsy remembers, "you couldn't bend over and tie your shoes without bumping your head on the dresser." There was no refrigerator; in summer they had to store milk by wrapping a cloth around it and putting it in a bucket of water. Eventually, they advanced enough to have an icebox, although one time Tombaugh slipped as he was trying to put some ice in it, damaging muscles in his abdomen. In some respects, Tombaugh said, they had "a feudal system at Lowell; we were the underdogs."

There was also a slight change in Tombaugh's relationship with V. M. Slipher. Although at this time the director still was doing some research, his administrative duties as well as his real estate business were gradually taking more of his time. He did continue with some important work, particularly in night sky emissions. Mounting a small spectrograph atop a thirty-foot tower near the 13-inch, he discovered some interesting ionospheric lines. Sometimes a difficulty arose when Slipher's real estate sideline interrupted observatory work. Often Tombaugh would have to leave his blinking to answer the telephone. He would mark his place in the comparator, leave his desk, answer the phone, get the director, return to his desk, and then try to remember if he had been moving across the plate to the left or right.

Tombaugh's university years were productive, and he naturally did well. In his senior year he did have one very difficult course on kinetic properties of gases ("I never worked so hard for two hours of credit!"). That year's study also included six hours of orbit computation, a challenging experience that increased Tombaugh's respect for the work that John Miller and the Sliphers had done for the first orbit of Pluto six years earlier. It involved getting positions for new comets and asteroids off the *Harvard Announcement Cards* and calculating their orbits by hand, using Carl Bremiker's tables of six-place logarithms.

Conversations and Mars

The cloudy or full Moon nights often continued to be interrupted by looks through the 24-inch or the 42-inch, or discussions with E. C. Slipher or Lampland about Lowell's legacy. A fierce supporter of Percival Lowell, Lampland always enjoyed telling stories about the observatory's founder and his work. Devoted as he was to Lowell and his theories of Mars, Lampland complained several times about Lowell's decision to build the administration building in a hollow rather than on a hill overlooking the town of Flagstaff. Lampland felt that the impressive architecture of the administration building would have been a beautiful sight from town. Instead, it was built for more practical reasons with the large refractor close by to its east. The large 42-inch reflector was to its west.

This large reflector was indirectly at fault for a disagreement between Tombaugh and the others that revolved around a telescope for planetary patrol. For years he tried to convince Lampland and the Sliphers that a well-made, long effective focal-length Cassegrain reflector could surpass a refractor in quality and would be ideal for a planetary patrol program. This effort was clearly an uphill battle. Although Lampland had used the 42-inch reflector successfully for many years, it did not have nearly the optical performance of the large refractor. Part of this reason was its strange mounting, dug into a pit below ground level. Originally, the idea was to lessen the temperature change from day to night by mounting the instrument below ground; the result instead was consistently poor seeing. Most telescopes today are constructed above, not below, ground level, to improve that aspect of their performance. Partly because of the way the 42-inch was set up, Tombaugh was the only staff member with unflagging confidence in the ability of a reflector telescope to give superior images. His efforts to get a large, long-focus reflector for Lowell's planetary program got nowhere.

E. C. Slipher was talkative and friendly. Unlike V. M., who was always working, E. C. enjoyed social activities like bridge, which his wife encouraged him to play. When the sky was too moonlit for planet search, Tombaugh would join him in observing planets. E. C. Slipher's drawings were beautiful and highly accurate, and he was a superb judge of proportion: "You know, Mr. Tombaugh, I have one personal equation I always have a problem with. I always draw the Sabaeus Sinus [a Martian feature] three-tenths of a millimeter too high." On some cloudy nights Tombaugh would listen to E. C. Slipher's stories, sometimes for an hour or more. As both men were interested in Mars, their talks continued and their friendship grew as the years went by. Tombaugh informally proposed a theory that the Martian canals existed along fracture or fault lines on the Martian surface, out of which warm vapor escaped to support a small amount of vegetation. In an article Slipher wrote for *National Geographic*, he attributed the fracture idea to Tombaugh.

While observing the planets through the 24-inch refractor, Tombaugh would often experiment with different eyepiece magnifications as well as stopping the lens down to various apertures. "When I used the same parameters that Lowell used, I saw the canals much as he drew them in the same geographical places; they were not figments of Lowell's imagination. I'll vouch for that; he was being honest with what he saw." The power Lowell used was too low, Tombaugh thought; the 400 power that he and E. C. used was much more revealing when the seeing conditions would permit such magnification.

So dedicated was Lowell to the idea that Mars was inhabited that he deliberately used low powers that made the canals appear straight. "He wasn't being scientific about it," Tombaugh notes. Also, Lowell's observations of canals were not restricted to Mars, as planetary observer Walter Haas points out: "Lowell drew canals on sketches of Venus, Mercury, and the Galilean Satellites of Jupiter. There is no question that he had a strong predilection to record features near the limit of visibility as narrow bands." [16]

Each albedo feature was a law unto itself, Tombaugh noted over years of Mars oppositions. Some of the maria would acquire a brownish tinge, but others would turn greenish. The canals showed enormous seasonal variations; some would appear at different seasons from others. Now we know the answer to be varying amounts of dust covering these areas, carried by the ferocious Martian winds. At the time, Tombaugh thought it was vegetation.

Although Lowell was not the first to study the "canals" on Mars, his thoughts were the most controversial. Lowell was entranced by Mars and its markings. "The lines," he wrote, "appear either absolutely straight from one end to the other, or curved in an equally appropriate manner. There is nothing haphazard in the look of any of them. Plotting upon a globe betrays them to be arcs of great circles almost invariably. . . . Their most instantly conspicuous characteristic is this hopeless lack of happy irregularity." [17] In his *Lowell and Mars*, William Hoyt points out that the canal controversy came up at a time when this planet was more interested in canal building than at any other period in its history.[18]

The Mars controversy may have reached its apex as a *reductio ad absurdum*, on October 31, 1938. As part of a weekly radio presentation by Orson Welles and his Fireside Theater, an hour-long drama based on H. G. Wells's *War of the Worlds* was broadcast on Halloween night. The drama was brilliantly produced to appear as an actual radio broadcast, beginning with a simulated musical presentation that was interrupted by reports of increasing severity about the landing of an army of Martians. The early report simply referred to an unusual happening in New Jersey, and a later one confirmed the alien nature of the invaders. Their origin was provided by an astronomer at a "Mt. Jennings Observatory" who had reported some curious light activity on Mars a few days earlier. The destruction the invaders wrought on the northeastern United States was complete, only to be followed by their own demise through viral infection.

The Tombaughs were at the University of Kansas that year. Like most listeners, Patsy had been paying only peripheral attention until the first report of a landing came in. Tombaugh was studying. "Come listen to this," Patsy said. "It sounds like people are coming from Mars."

"What?" Tombaugh was incredulous as he went toward the radio and heard one of the early announcements. "Boy, how fortunate, here we are experiencing an invasion from Mars!" Tombaugh recollects his somewhat naive thought:

> What a superb, what a monumental event this is. That's the way I felt. Here in my lifetime, this is the first time this has ever happened! Of course, at that time I had halfway—but only halfway—believed that Mars had intelligent beings. You know how dramatic Orson Welles was; it seemed like a most remarkable event. It took us completely by surprise. Everything sounded so realistic. I was amazed; I thought, how thrilling this is! Come what may, I was delighted because maybe Mars was inhabited; the skeptics would sure have to face it now. Then they said they had seen a light from Mars a couple of nights earlier. The jig was up for me! I knew at that time Mars was squarely behind the Sun. I was halfway disgusted and halfway amused.[19]

Unfortunately, most of the station's listeners did not share Tombaugh's knowledge that Mars was unobservable, as well as his suspicion that the Mt. Jennings Observatory did not exist. As the hour progressed, more and more people tuned their radios to the network, and by the end the nation was in a state of panic. A coincidental power failure in the state of Washington convinced many people that the danger was real.

Tombaugh admits that had the sightings of the launch from Mars not been mentioned so early, he might have gone on believing that the invasion was real. The panic was international. While visiting Quito, Ecuador, in 1956, he learned that the rage was so great that "people had wrecked the radio station over it. They were panic-stricken, then became infuriated and just wrecked it!"

One can rarely predict public reaction to a chain of events. The Mars controversies had been simmering for decades, but obviously Lowell's idea of intelligent life on other planets had succeeded in striking a powerful chord that sounded loudly that night in 1938. The Welles drama also had the advantage of occurring at the height of a long period of great public trust in science that had begun, Tombaugh believes, during the Scopes monkey trial of 1925. After a peak of confidence during the atomic bomb effort of the Second World War, public perception of scientists began to drop during the "mad scientist" fifties. The October 1957 launch of Sputnik "changed everything by 180 degrees; suddenly the scientists were heroes, and a program was immediately instituted in the schools with an enormous emphasis on mathematics and science."

Chapter 7

GALAXIES

NOT LONG after he started blinking in 1929, Tombaugh decided to keep a count of the numbers of objects of different types. Each plate's variable stars, asteroids, comets, and galaxies gave him a good idea of what the sky had to offer. There was even a category of "temporary objects," including suddenly rising variable stars or possible novae, in which an object is apparent on only one of three plates. Having begun the listing, he realized that it entailed relatively little additional work, and he continued it throughout the entire fourteen years of work. "My thought was to get everything out of this tedious blinking that I could. If I were going to all this difficult trouble, bleed it to death! That was my attitude." The wisdom of this idea was never more apparent than with the distribution of galaxies, a subject that was becoming increasingly significant.

Although this approach had an observational basis, it included a strong theoretical component. Trying to keep up with the discoveries being made at Mt. Wilson and other observatories around the world was difficult, but Tombaugh did read nearly every journal that came in to the observatory. "I was quickly going through the journals, trying to keep up on astronomy, not just planets but stellar and galaxies; it made me much more alert when I ran into something."

The Nature of Galaxies

A century and a half earlier, planet discoverer William Herschel had started to wonder about the strange nebulae that his telescope was

105

picking up in addition to the planets, stars, and clusters of stars he was seeing. Curious fuzzy patches scattered over much of the sky, the nebulae were especially puzzling. Already our understanding of their nature had undergone several changes; in 1755 the philosopher Immanuel Kant had proposed that nebulae were island universes, each one being equivalent in size and complexity to our own Milky Way.[1] Although Herschel accepted the island universe theory at first, by 1791 he was changing his views about at least some nebulae that had starry central condensations. After all, if we could see these stars in other universes, they must be inordinately large. But another aspect of the nebulae was bothering him. How could they be independent of our Milky Way system if their distribution seemed to be wholly related to it? Nebulae that could be resolved into stars abounded in the plane of the Milky Way, yet nebulae that could not be so resolved tended to concentrate away from it. To Herschel it was obvious that if nebulae were really separate universes, their distribution would not be so intertwined with our own system.[2]

Early this century our understanding of the huge systems of suns that we call galaxies took a leap forward at Lowell Observatory. Acting on a suggestion by Percival Lowell, who suspected that fuzzy, nonstellar objects called *spiral nebulae* might be new solar systems being formed, in 1912 V. M. Slipher took a spectrum of M31, the first of more than forty spectra (at exposures of up to forty hours each!) of different nebulae. He noted that these spectra showed anomalous shifts to the red, a discovery that later indicated that the nebulae must be moving away from us. In one astonishing case Slipher measured a galaxy in Virgo to be receding at a velocity of eighteen hundred kilometers per second.

Slipher's find probably did more than anything else in a century to push ahead the understanding of nebulae. Late in the nineteenth century Rev. T. W. Webb wrote that the Andromeda Nebula, one of the finest in the sky, "*may* be an exterior galaxy."[3] But the spark that really set off the new thinking was ignited in 1920. In what has become known as the Great Debate, Harlow Shapley, of Mt. Wilson, and Heber Curtis, of Lick, met at the National Academy of Sciences in Washington on April 26, 1920, to discuss nothing less than "the scale of the Universe." Shifting the emphasis from scale to the nature of the spiral nebulae, Curtis claimed that these nebulae were outside our own system. Citing other evidence, Shapley disagreed.[4] Although Shapley was clearly wrong here, he was closer to being correct in his evaluation of the size of the Universe, using the distance he measured to the globular clusters (see Chapter 3).

Four years later, using the 100-inch reflector on Mt. Wilson, Edwin Powell Hubble used Cepheid variable stars to determine the distances of many galaxies, following the technique used by Shapley with globular clusters. Hubble was instrumental in determining that the spiral nebulae were very distant masses of stars. His counts of "extragalactic nebulae" produced basic information on our understanding of how the light from galaxies is obscured by dark matter in our own Galaxy. As the distant nature of these objects became better understood, Shapley suggested that calling them *nebulae* would confuse them with the gas clouds of our own Milky Way and that they should instead be called *galaxies*.

The Perseus-Andromeda Stratum

After his graduation from the University of Kansas in the spring of 1936, Tombaugh was back at Lowell, blinking plates again by August 12. This period lasted uninterrupted until June 30, 1943, with the exception of his 1938–39 master's degree year. He arbitrarily dropped the "X" numbers that the other observers had used to organize plates and resumed the count with "1100" numbers. By now the blinking was so far behind that Putnam and Slipher obtained two small grants from the American Philosophical Society's Penrose Fund so that the search could proceed faster. Henry Giclas then took over the photography so that Tombaugh could spend all his time blinking.

His search began from the zodiac, and a series of regions in Pegasus, Andromeda, and Perseus was photographed. Soon Tombaugh noticed a sudden change—a major increase—in the numbers of "extragalactic nebulae" (now known by Shapley's term of *galaxies*) he was counting in a plate involving the northern half of the Great Square of Pegasus. The density of galaxies had risen to about six per square degree, about twice the typical rate for areas away from the galactic plane. Taking a week to nine days to blink through a single pair in these regions, it was several months, well into 1937, before he realized the extent of what he had found.

What he had revealed was a vast cloud of galaxies beginning suddenly in the Square, passing behind Alpha Andromedae, proceeding eastward along an arc south of Beta and Gamma Andromedae (including northern Triangulum), contracting sharply in Perseus as its northern part met the Milky Way, and reaching a sharp point and stopping just east of Beta Persei. So large was the clustering that at first Tombaugh was not aware that what he had seen was extremely unusual; he may have gone through an entire pair of plates before he noticed that this was more than just a chance increase in the numbers of galaxies. Through plate after plate this clustering progressed; "I was really beginning to get flabbergasted about it. I'd mark on the plates where the density would seem to fall off and then plotted on an atlas the shape and extent." The edge of the group in Perseus was so pointed and obvious that it removed any doubt in Tombaugh's mind that it represented an organized system.

The brightest of the member galaxies were of the fourteenth magnitude, and he detected them almost to the telescope's limit of about seventeenth magnitude. Not only was the sheer size of this system unusual, but so was its shape, a bit like a banana. The pointed eastern edge could be explained partly by interference from the Milky Way, but there seemed to be no ready interpretation for the sharpness of the rest of its boundary.

Unlike the other discoveries, this one was gradual, so the director was not surprised by the time Tombaugh spoke to him about its extent. "I think we have a very interesting concentration of galaxies here that has exceeded anything I've run into," Tombaugh concluded. Although this one was the largest, it was not the only cluster of galaxies discovered during the search.

He called this huge mass the Great Perseus-Andromeda Stratum of Extra-Galactic Nebulae. Director Slipher was delighted with Tombaugh's find and was anxious that it be published. Tombaugh suggests that Slipher was "particularly proud that the Lowell Observatory had found that." Submitting the paper to the *Publications of the Astronomical Society of the Pacific* (*PASP*) was Tombaugh's idea; at the time it was one of the most widely read astronomical journals. It was surely the one Tombaugh opened most enthusiastically when it arrived at Lowell's library, for it contained the latest news of what was being discovered at Lick and Mt. Wilson.

Accordingly, Tombaugh prepared a paper for the June 1937 meeting of the Astronomical Society of the Pacific in Denver and for later publication in *PASP*. This paper shows the enormous extent of the stratum; in all, Tombaugh counted some eighteen hundred galaxies in it.

> The apparent shape of this cloud of nebulae is that of a great stratum curved somewhat like an arc, the concavity facing the Milky Way to the north. It is about 45 degrees long and from 5 to 10 degrees wide. The widest parts are two great lobes, one at each end. Most of the cloud is situated in the southeastern part of the constellation of Andromeda, the remainder in the northern parts of Pisces and Triangulum, and western Perseus. The star Alpha Andromedae marks the center of the rich lobe at the west end of the Great Stratum. This lobe itself is elongated in an east-northeast-west-southwest direction, extending about 8 degrees on each side of Alpha Andromedae. The east end of the Great Stratum is marked by the well-known variable star Algol.[5]

Unnoticed by anyone on the Lowell staff until Brian Skiff pointed it out to the author in the mid-1980s, the huge clustering had actually been reported earlier by Walter E. Bernheimer, at Lund Observatory, and K. Lundmark,[6] and was mentioned in a 1932 issue of *Nature*. In a footnote to his *PASP* paper Tombaugh admits that Mt. Wilson astronomer Fritz Zwicky had earlier reported a cluster of nebulae in Pisces which now appeared as a small section that "corresponds to the west central part" of his greater "congregation of nebulae."[7] It was Zwicky, in fact, who first demonstrated, using the 18-inch Schmidt camera on Palomar Mountain, that clusters of galaxies were common in space.

Regardless of who originally identified the system, its full dimensions were first revealed on Tombaugh's 13-inch plates. "I found clusters within the cluster, and then clusters within a cluster, within the cluster. It was the richest region of galaxies I had ever encountered." The harvest that Tombaugh counted on all his plates fascinated him in both appearance and distribution. The *RN* designations for "remarkable nebulae" dot many plates, as do *DN* markings for "double nebulae." In the twilight of the thirties Tombaugh was making plans that unhappily would never be fulfilled: "When I got the blinking done—I had to do it quickly before [the plates] got too old—I was going to take a few years working up all the by-products of stuff I'd seen and recorded on the plates." He was especially interested in studying the distribution of the unusual nebulae and the double nebulae.

The occurrence of double nebulae, which he defined as two galaxies within three diameters of the larger member, was much more frequent than he had been led to believe from his reading of the literature. If a pair is the same distance from us as well, rather than simply two galaxies that happen to lie on our line of sight, it becomes likely that they are interacting physically. The only way to determine their distances was by measuring the shift of the red end of their spectra; the larger the "red shift" of a galaxy, the greater the velocity at which it recedes from our point of view. Taking spectra was not part of Tombaugh's planet search, but in a plate survey that covered so much of the sky visible from Flagstaff he counted many of these possible interacting galaxies.

The Problem of Galaxy Distribution

Farsighted and imaginative, Edwin Hubble had also been taking a survey of the arrangement of the galaxies through space, avoiding both the Milky Way (an area he called the "zone of avoidance") and the obvious clusters of galaxies. His famous book, *The Realm of the Nebulae*, was widely read by amateur and professional astronomers, including Tombaugh, who called it "one of the great books of the time." It commented on galaxies' distribution:

> The agreement between the polar caps, together with the absence of systematic variations either in longitude or in latitude, may be summed up in the statement that the large-scale distribution over the sky is approximately uniform. This conclusion is drawn from only a fraction of the sky. The zone of avoidance with its bordering fringes withdraws large areas from possible exploration and, in addition, about 25 per cent of the sky cannot be effectively observed from the station at which the survey was made. However, the areas investigated include both galactic poles, the whole of the northern cap, 60 per cent of the southern cap, and the less obscured portions of perhaps two thirds of the galactic belt. The extent and the pattern of these regions would appear to constitute a fair sample of the sky as a whole, and the complete absence of

appreciable, systematic variations strongly suggests that no significant, major departures from isotropy [having similar physical properties in every direction] should be expected in the unobserved regions.[8]

Although the 100-inch reflector could penetrate to a very faint magnitude limit, it could not cover the entire sky. To bypass this difficulty, Hubble took plates at specific and evenly distributed regions of sky. Besides the paucity of galaxies around the zone of avoidance of the Milky Way, the farther he moved from it the more galaxies tended to appear, and the result of his survey showed an isotropic, or even, distribution; counts from his specific plate regions clearly indicated that galaxies are distributed evenly. Derived from these findings was his "cosmological principle" that the uniformity we see really portrays the structure of the Universe.

Hubble was aware that on relatively small scales, for instance in the constellations of Coma Berenices and Virgo, the distribution "is conspicuously nonuniform. Nebulae are found both singly and in groups of various sizes up to the occasional, great compact clusters of several hundred members each."[9] To avoid having these clusterings bias his results, he omitted them from his survey; his point was that over the total volume of the Universe, the distribution was even. He also suggested that all galaxies may be clustered, but that the clusters were evenly spread. Large as the Coma-Virgo system appears from here, on the scale of the entire Universe it is just one of thousands of clusters of similar size; it is not that large and would not be visible as a beacon from the Universe's most distant reaches.

Hubble thought that his cosmological principle applied at a level above that of the clusters of galaxies. Tombaugh disagreed. With discoveries like his huge stratum with its sublayers of clustering, he felt that the distribution might be nonuniform at this higher level. It was the scale of uniformity at the cluster level that caused Tombaugh to dispute Hubble. During all this time Tombaugh was a planetary astronomer (an appellation, he notes, then looked down on by other astronomers) who took wide-field plates of large areas of sky. He began to notice how uneven the distribution of galaxies was, even in areas of the sky far from the equator of our own Galaxy, whose obscuring dark matter would block the light from other galaxies. At lower galactic latitudes Tombaugh checked nebula density against possible Milky Way contamination. "In my examination of plates this summer," Tombaugh wrote on the envelope for plate 487 around August 24, 1933, "there was noticed a marked decrease in nebulae on the pair just to the east of this one, and the decrease on this pair is still more pronounced; on the other hand, this pair of plates has shown a noticeable increase in the number of stars. In other words, the effects of the Milky Way are showing." [10] At high galactic latitudes the density of galaxies ranged from a high of thirty-three and a half galaxies to a low of one-third galaxy per square degree.

By keeping careful notes about the numbers of galaxies on each of his plates, Tombaugh was conducting a survey of galaxies down to about sixteenth magnitude from the entire sky visible from Lowell Observatory. It was not the first survey, but its importance has been very much understated. Early surveys by Charles Messier (about 110 objects), William Herschel, and John Dreyer (his *New General Catalogue*, mentioned in Chapter 6, contains more than 7,000 objects) vastly increased our understanding of the distribution of nonstellar objects by the end of the nineteenth century, but these surveys were more thoroughly done in some regions of the sky than others.

Possibly the first all-sky survey intended especially for galaxies was prepared from 1930 to 1932 by Harlow Shapley and Adelaide Ames from Harvard College Observatory patrol plates. The importance of this survey is that it took place just a few years after the discovery that galaxies were distant objects, far removed from our own Milky Way. To about magnitude thirteen, 1,249 galaxies were listed—not a high number, but it did have the advantage of being taken from plates uniformly exposed across the sky. At the time, Shapley noticed that the distribution was spotty and that galaxies were clustered.

Tombaugh's trans-Saturnian planet survey produced a total of 29,548 galaxies. Later surveys by Fritz Zwicky, of the California Institute of Technology, produced more than 30,000 galaxies, as well as 10,000 clusters of galaxies of various sizes, and a listing by Vorontsov-Velyaminov, of the Sternberg State Astronomical Institute, contains almost 35,000 objects. [11]

In 1945–46 Tombaugh was teaching a course at the University of California at Los Angeles and had the chance to meet with Hubble several times by driving to his Pasadena office. From Tombaugh's viewpoint, the two men were seeing things from totally different directions. By that time Tombaugh had examined about 70 percent of the sky to a resolution of four arc seconds and a magnitude limit of about seventeen. "I knew that the distribution [of galaxies] was extremely patchy, but I couldn't convince him." With the mighty 100-inch, Hubble had seen much smaller areas, but to a far dimmer magnitude limit than Tombaugh could do with a 13-inch camera. Tombaugh recollects one of his 1946 conversations: "I began, 'Dr. Hubble, on my plates, I have made counts. I have seen the area, and I don't agree with the conclusion you have drawn in the distribution. I find marked irregularities. I see the voids and the concentrations, even at high galactic latitudes. What I have seen on my plates does not agree with what you have said.' He seemed a bit shocked about it. But I couldn't convince him." [12]

Hubble answered merely that the plates he had seen, taken in small areas but over selected regions spread out over the sky, did show a fairly uniform distribution. Tombaugh then offered to show him his records, in which he had kept careful counts of the number of galaxies in each plate. Hubble was apparently not interested in those either. It would have involved a lot of looking on Hubble's part through the counts of 29,548 galaxies. Probably, Hubble did not realize how much material Tombaugh had assembled. Never before had such a detailed survey of the sky been taken, and another one would not begin for some years. It is also possible that Tombaugh's lack of a doctorate was one reason that his arguments were not taken very seriously.

In 1937, the year of Tombaugh's paper in *PASP*, Erik Holmberg and Anders Reiz independently demonstrated that the Local Group, including our Galaxy, is a part of a much larger swarm of galaxies. Two decades later Gerard de Vaucouleurs began a much more intense study that revealed the existence of the Local Supercluster, which includes both the Local Group and the Coma Berenices and Virgo clusters. Considering Tombaugh's own experiences, it might not be surprising that many astronomers, including, apparently, Edwin Hubble, ignored these two important pieces of work.

In the second half of the 1940s, photographic surveys of the entire sky were begun at Lick Observatory and at Palomar Mountain in California. The Palomar Survey, funded by the National Geographic Society, produced the most extensive atlas ever of the entire sky. (It did not have to be redone until the latter half of the 1980s, when better photographic plate emulsions and a technique called *hypersensitizing* were available, in which plates are treated with a "forming gas" of nitrogen and hydrogen to increase their sensitivity.) George O. Abell, a central figure in the early project, catalogued 2,712 clusters of galaxies in the sky that could be observed from Palomar. In 1958 he published his results in a huge Ph.D. dissertation demonstrating that several thousand large clusters observable at very large distances are not randomly distributed but tend to associate in huge "great clusters," or superclusters.

That was exactly the point Tombaugh had been trying to make to Hubble thirteen years earlier. Abell concluded from his and other studies that the matter in the Universe seems to be gathered into superclusters some three hundred million light-years across. Although Hubble's cosmological principle of uniform distribution may be maintained at the supercluster level, on a smaller scale there are the voids and concentrations of clusters into superclusters that Tombaugh's plates showed so well. In addition to the supercluster Tombaugh recorded, and the Local Supercluster that includes our Local Group and the Virgo cluster, other superclusters were eventually found centered in the constellations of Fornax, Hydra, and Pavo.

Above the scale of the superclusters, does the Universe continue to be hierarchical? Are there megaclusters, or clusters of superclusters, and clusters of megaclusters? The answer is important, since the average concentration of matter in the Universe is much reduced from what it would be according to the cosmological principle; in fact, if the hierarchy goes infinitely, the average density would approach zero! Abell predicted that there is no hierarchy beyond that of the supercluster.[13]

Chapter 8

WAR AND DEPARTURE

I N 1938 Tombaugh decided to return to school for his master's degree. He had considered one of the schools in California but found them too expensive, and even though conventional wisdom suggests that one's graduate degree be earned at a different school from one's first degree, finances dictated he return to the University of Kansas. As Patsy began her senior year that fall, Tombaugh returned to Kansas to begin work. Dinsmore Alter was no longer there; he had left in Tombaugh's junior undergraduate year for the Griffith Observatory in California. Norman Storer was the astrophysicist who supervised Tombaugh's thesis, a study of the observational capabilities of the university's 27-inch Newtonian reflector, coupled to a program to restore the telescope to pristine condition.

Arriving at that peculiar subject was the result of two types of thinking. The first was practical; Tombaugh reminisces that "somebody had to get the telescope ready to use," although telescope construction and maintenance is normally a subject for a telescope-making firm, not an academic degree. The second reason was ethereal; Tombaugh had "a profound love for working with telescopes." The combination of lines of thought was sufficient to turn a mundane subject into an inspiring piece of scholarship that combined mechanical aptitude with recording and evaluating the work on paper and producing a master's thesis.

When Tombaugh arrived in Kansas, the telescope was useful only for visual work and for the occasional Moon photograph that Patsy's brother James Edson had taken. After Tombaugh had completed work on the instrument, he took its first long-exposure photographs. The guide telescope was fashioned from one of the 12-inch mirrors Tombaugh had ground earlier. For prime focus imaging, he prepared an instrument that enabled focusing to 1/200 inch. It involved one of his old opera glasses taken apart and installed to project a star image off to the side to an eyepiece. In effect, one guided on "the image of a star image." In 1980 the observatory housing this telescope was named for Clyde Tombaugh.

The graduation ceremony of June 1939 featured two Tombaugh degrees: Clyde received his master's, and Patsy her bachelor's degree. He never continued to a doctorate. After his master's he did consider it, but he did not have any money for the minimum three-year undertaking. Moreover, "part of the reason to get the doctorate," Tombaugh thought, "was to get a union card. But I already had made my reputation on my own; that was one reason for finally abandoning the idea." At Lowell he felt established and comfortable; both E. C. Slipher and Carl Lampland had only honorary doctorates. Feeling he had learned much more about planets at Lowell than he could at a university where doctoral programs in the planetary sciences were rare, he felt no pressure or desire to go further. "I was established; I had a place and a position." In any event, in 1960 he received an honorary doctorate from Northern Arizona University.

After graduation the Tombaughs returned again to Flagstaff. With funds no longer available from the two grants from the American Philosophical Society, Tombaugh began plate taking again.

Schmidt Cameras

Had the observatory ordered its planet-search camera in 1930 instead of 1927, its staff would probably have decided on a Schmidt camera instead of the astrograph, a decision that would have had consequences in shorter exposures and plates sensitive in red as well as blue light. For in 1930 Bernhard Schmidt of the Bergedorff Observatory near Hamburg constructed a photographic camera with a spherical mirror at the back and a thin lens at the front. Spherical mirrors suffer from unsharp images because their edges do not have exactly the same focal length as their centers. The thin correcting lens adjusts this problem, with the result that images remain sharp over a very large field of view.

When the Lowell staff found out about the Schmidt camera, they were incredulous. Lampland discussed the news of the Schmidt with Tombaugh, and both were "agog" to learn of a reflecting system that would have given much better red sensitivity, preferable to the refracting system they had, which favored blue light. Also, the system would have been much faster, meaning that exposures could have taken a fraction of the time the astrograph needed. Blinking, of course, would have taken exactly the same amount of time.

The next chance for a Schmidt came around 1941, when enthusiasm started to grow for a search down to a fainter magnitude. Tombaugh proposed an f/4 system for its combination of long exposures without fogging of the plates by the sky background. Although Slipher originally had suggested a faster system, Tombaugh prevailed and the glass was ordered. The next step was to take a series of two-and-a-half-hour exposures around the zodiac with the existing 13-inch to compare the results with the one-hour exposures. He began in Pisces, continuing into Aries and Taurus, repeating (but to almost a magnitude deeper) the areas he had photographed back in the autumn of 1929. Part of the reason to change the strategy was that the one-hour exposures had been taken far from the ecliptic, where the chances for new planets were getting scarcer and scarcer; he was not even seeing any asteroids! Alert to any anomaly, he made the interesting observation that on the new plates, the asteroid trails showed variations of the order of a magnitude.[1] Since some asteroids rotate in periods of a few hours, he could learn something about this aspect of their behavior just by studying their trails.

Meanwhile, the onset of World War II had thrown the future of the planet search into doubt. If the U.S. had to enter the war, this work would undoubtedly stop. Reasoning that in the time left it made more sense to search to a fainter limit but closer to the ecliptic, Tombaugh began the new strategy. It resulted in some difficult examinations, especially in the rich areas of Sagittarius, where he needed almost a month of work to blink a single pair of plates!

On one bitterly cold winter night Tombaugh was photographing an area near Messier 44, the beehive cluster, in Cancer. Pluto was there at the time, and Tombaugh was searching for a companion of his planet. Dressed in a heavy sheepskin coat, sweater, and felt-lined boots, he had thought he was warm enough as he started the exposure around ten o'clock that evening, just after moonset. The telescope was tracking so well that Tombaugh could watch the guide star sitting in the center of the eyepiece, not moving at all. Sometimes he could go for fifteen minutes or more without making an adjustment. Thinking how fortunate he was to have such a well-mounted camera, he relaxed, growing strangely comfortable and sleepy. He even thought of lying on a Navajo rug on the floor to nap for a few minutes.

About two hours into the exposure, with just half an hour to go, it was time to make a small adjustment; he was about to reach for the paddle. But where were his fingers? They would not move; he could barely feel them. Suddenly, he knew his comfortable feeling was an early stage of hypothermia. He forgot about finishing the exposure; at 2.25 hours the plate would be fine as a third plate to check any suspects. Closing the telescope and dome shutters was agonizing, and covering the plate with a dark slide, normally the work of a few seconds, took five minutes. He succeeded only by putting his fingers on his head from time to time, underneath his stocking cap.

Never had the short three-hundred-yard walk back to the administration building been so difficult. His hands were numb, and he was barely able to move his legs. When he finally got to the building, he struggled to get the plates safely into the darkroom and finally made it to the radiator on the second floor. As he stood, feeling began to return to his extremities, and with the feeling came waves of pain. He stayed by the radiator for more than an hour.

Despite the difficulty in blinking plates that extended the planet search to magnitude eighteen over 22 percent of the zodiac,[2] the observatory decided to acquire a 25-inch f/4 Schmidt camera. With such an instrument, regions down to nineteenth magnitude could be searched, a limit 2.5 times fainter than the long exposures with the 13-inch and more than six times fainter than the limit of the conventional one-hour exposures.[3] The outbreak of war put the Schmidt plans on hold, and the expansion of the planet search was not resumed afterward. As late as 1989, plans were still afoot to obtain a large Schmidt camera for Lowell Observatory. The glass still remains.

The War

On the night of December 7, 1941, Tombaugh was slowly grinding a new 16-inch blank when Carl Lampland came down, having just heard some news over the radio while at home eating dinner. "You know, Mr. Tombaugh, the Japanese have just bombed Pearl Harbor." Those words meant the beginning of a drastic change at Lowell, although it did not occur suddenly. The ponderosas and snows of the observatory were not enough to shelter it from the upheaval that the war would bring. Tombaugh remembers the war's "desperate struggle; we were not even sure we would win. I lost a lot of sleep thinking how Hitler was razing those countries; I thought the world was doomed. I was very depressed about it."

Tombaugh continued making mirrors during the war years. In 1944 he finished a 10-inch, a telescope that years later would watch the stars on a Dobsonian-type mount modified to sit on a discarded lawn-mower chassis! The azimuth motion would sit on an 8-inch-diameter aluminum disc riding against a plate, also made of aluminum, and bolted onto the bottom of the Dobsonian fork.

Genuinely fearful that the United States faced a "clear and present danger" from Germany and Japan, in July 1942 Tombaugh joined a one-week class in civil defense. It taught how to recognize shells of "dud" bombs, how to identify properties of various poisonous chemicals, how to organize an air raid system, and where to put large numbers of people if an emergency occurred. One Sunday morning after the class ended, Tombaugh had a visitor named William McKee. "We want you to be the commander of civil defense of Coconino County," he announced. Apparently, of all the people taking the course, Tombaugh had not unexpectedly scored by far the highest.

The attack at Pearl Harbor had crippled the U.S. fleet more seriously than the Japanese probably realized, and there was concern about Japanese infiltrators setting forest fires in the heavily wooded county. What would happen if severe fires caused one hundred thousand refugees each day to pour into Flagstaff? Tombaugh gave lectures between movies at the same Orpheum Theater that had shown *The Virginian* the night he discovered Pluto. During these brief lectures he explained some of the possibilities that a civil defense emergency might entail and asked for volunteer firefighters, air-raid wardens, and nurses' aides. In

his usual direct manner he must have been quite persuasive in attracting volunteers, warning that "if we get attacked, you will be the victims!"

For the first two months of the program Tombaugh was involved several hours each day, in addition to his regular observatory duties. V. M. Slipher fully supported Tombaugh's civil defense work, allowing him the necessary hours to do it. "I was busier than a cranberry merchant at Thanksgiving," Clyde said. Patsy assisted as part of a team counting the numbers of cots, blankets, shoes, and other merchandise available in county stores for quick use. At one point Tombaugh traveled to Phoenix to learn about the threat of "personnel bombs" attached to weather balloons. Although some of these weapons flew as far east as the Great Lakes, none traveled as far south as Flagstaff. It turned out that the only death that occurred in the continental United States as a direct result of enemy attack was due to one of those balloons, which landed in the state of Washington.

In February 1943 Tombaugh, still working full time at the observatory, was invited to teach physics at Arizona State Teachers College in Flagstaff. The regular teacher had left unexpectedly, and the class was in danger of losing its year if a substitute could not be hired. Thus Tombaugh found himself a professor of physics.

That same year, at age thirty-seven, Tombaugh faced military service. For a brief period he expected to be serving overseas, but one day he got a message to report to a navy commander. Wondering what he could possibly have done wrong, he reported. Tombaugh reconstructs the conversation:

"Mr. Tombaugh, we want you to teach navigation."

"But I've never had any navigation!" said the retired Kansas farmer.

"Oh, but I've been looking into your background, and you're the one that is best suited for it." At least Tombaugh did have the astronomical and trigonometric background.

"I thought someone with more mathematics would be teaching it." Up to then the bulk of Tombaugh's navigation experience had been limited to miniature boats on his farm, but he did know that there was a scarcity of people qualified to teach the subject. However, he was still occupied teaching physics.

"We'll relieve you of physics; we want you in navigation."

"But I understand you do not even have a textbook!"

"We have a copy of Bowditch," the commander explained, referring to a well-known standard in the field written by Nathaniel Bowditch (1773–1838), first published in 1802, and still widely used.[4] "Until the books come in, you can lecture from this. Good luck!"

The draft board agreed to exempt him from service so that he could teach navigation for the navy at Arizona State Teachers' College. Students came from all over the country, and for the first of his seven semesters of teaching the course he had to work virtually around the clock just to keep ahead of them. Reminiscing that at first he "didn't know port from starboard," he was a fast learner and soon became familiar with lighthouses, buoys, and other aspects of coastal navigation. The second semester, celestial navigation, was far more comfortable for him, and he brought all his experience as an observer and planet hunter to bear on his new assignment.

Tombaugh approached this task with his customary thoroughness, but in this case he had no choice. Students who took the class then boarded navy ships and went into combat. The thought that a navigation error could lead to disaster made both lecturer and students approach their work with care. He thought he was doing it right, that he understood the concepts he was teaching, but could he actually figure out where he was?

He didn't know the answer until a real-life experience came years later during military flights from Aberdeen Proving Ground in Maryland to White Sands, New Mexico. Tombaugh tried to keep a reckoning of the plane's position, using a watch, a compass, a road map, and a ruler. Occasionally, the pilot asked him if he knew where they were, and he answered. On one journey, above the clouds, Tombaugh was confused because his observations indicated that they did not appear to be taking the route he had expected. "It's funny," he told the pilot, "but it looks as though we should be right over Charleston, West Virginia."

"We're right over it!" was the reply. If Tombaugh had taught his students as well as he had taught himself, they would have been all right.

Teaching navigation virtually stopped his planet search. Because he was no longer working full time at Lowell, he did not expect to be paid; nevertheless, at first he still did some blinking on weekends. One afternoon V. M. Slipher walked by, somewhat surprised to hear the comparator running. "Clyde, why are you blinking?"

Tombaugh looked up and answered, "Because I think these should be blinked, and I don't know what's going to happen in the future." Not long afterward the director gave Tombaugh a check for this extra work.

With a down payment from money that Patsy had recently received from the estate of her grandfather, the Tombaughs finally acquired a house in town and moved out of their cramped quarters at the observatory. It was a practical move to find better living quarters, but V. M. Slipher and some others "may have felt that it broke with tradition."

Departure

As the war years progressed, Tombaugh became troubled by a growing feeling that V. M. Slipher was unhappy with him. Slipher's treatment of the Tombaughs had begun to deteriorate late in the thirties. Patsy remembers a strange incident involving Slipher's asking if she would clean a room for an astronomer visiting the observatory. As she was cleaning, Slipher and the astronomer walked into the room. He never introduced the wife of the discoverer of Pluto to this astronomer. Instead, in Patsy's presence Slipher turned to him and said, "I'm sorry I didn't get you a maid before this." The growing evidence of this treatment was underscored when Patsy learned one afternoon that Slipher had become bothered that it was Tombaugh and not he who had found Pluto.

This resentment did not, of course, excuse Slipher's treatment of Tombaugh, but the director may have had a point. He had had a leading role in a planet-search program that had been running for about a quarter of a century, and Tombaugh had not even appeared on the scene until the final year before the discovery. To Slipher, credit for the discovery should have been shared among everyone who had participated in the project, and all these years later he remembered Tombaugh still as a young man from Kansas who was following orders. If that was his approach, he would naturally be disturbed that so much of the credit had landed on Tombaugh.

Conscious or not, V. M. Slipher's attitude was distressing to Tombaugh, and another incident underscored his unhappiness. In 1936 asteroid 2101 Adonis was discovered by Delporte a few days after it had approached to within 2.4 million kilometers of Earth.[5] In 1938, in response to a request for a search, Tombaugh worked with Henry Giclas to attempt to recover that object. They took a series of plates guided not at the normal sidereal rate of stars moving across the sky, but at a rate matching the motion of the asteroid; the stars would trail slightly but the asteroid would not. Although the search was unsuccessful, probably because of the asteroid's faintness at the time, Tombaugh and Giclas prepared a paper explaining their efforts. "Slipher put it in a drawer," Tombaugh recalls. "I think the thing that bothered Slipher was that he didn't want the young men publishing papers when the older ones weren't publishing any." It is also possible that Slipher simply did not want the paper published because the search had been unsuccessful. In any event, Adonis was not recovered until 1977, when Charles Kowal found it using the 48-inch Schmidt at Palomar and a prediction by Brian Marsden.[6]

In the spring of 1944 Frederick Leonard, a highly regarded teacher of astronomy at the University of California at Los Angeles, invited Tombaugh to teach two semesters there. A specialist in meteorites, Leonard had known Tombaugh since the early thirties. Tombaugh's courses were beginning astronomy and history of astronomy. It was a different experience from the quiet at Lowell and very hectic. Between the teaching and the grading of papers, there was virtually no free time or observing except for occasional views through a 5-inch refractor on the roof. Tombaugh felt that the long lines of students snaking up the narrow and winding staircase to the telescope was an inefficient way to introduce a group to the stars, so he later set up his own 10-inch reflector on campus.

On his return from Los Angeles, Slipher summoned Tombaugh and told him that the observatory finances were very tight and that he "should find other employment." Tombaugh was so shocked that he does not recall the exact conversation. He does remember that Slipher pointed out that given the physics and navigation teaching Tombaugh had done, and especially the year he had just completed in California, his real calling lay in education and he should try to find a position teaching. "He was so hurt he didn't even tell me," Patsy notes, adding that it was many months before she learned of the nature of that fateful conversation.

Part of the reason for change was obviously financial. Tombaugh had been away teaching; not having to pay him was expedient for Lowell's always meager finances. Also, with the planet search winding down, the observatory may have felt that Tombaugh was simply no longer needed there. As soon as V. M. Slipher had suggested he leave, Tombaugh lost all interest in doing anything more with the plates. The hurt feelings were so deep that he did not bother debating with Slipher about the importance of continuing with the studies of galaxy distribution and other by-products of his planet search. It was never Tombaugh's nature to insist on continuing when he felt that he wasn't wanted, even in the unlikely event he had the means to do the work independently. "I felt I had earned a status of some import and they weren't going to give it to me. . . . I felt terribly upset to know that all those data on nebulae were down the drain. . . . I could have made a real name for myself in the galaxy studies. I felt terribly cheated. I had all these years of hard work, hard blinking."[7]

By the end of the war V. M. Slipher's interests had turned away from basic research, and it is entirely possible that the director did not realize the importance of the galaxy statistics that had been assembled. Tombaugh repeatedly emphasizes that the war changed everything and that after his physics and navigation teaching, and his year at UCLA, the Lowell environment somehow became so hostile that the possibility of further galaxy work was shortchanged. By 1945 Clyde and Patsy sensed that Slipher was showing feelings of jealousy toward Tombaugh. It is likely that a combination of Slipher's uncertain feelings about Tombaugh's recognition for the discovery of Pluto, and Tombaugh's lack of confidence in proposing his own future direction at the observatory, led to this unhappy result. Whatever Slipher's reason, his decision to ask Tombaugh to leave meant that the important work on galaxy distribution was never done at Lowell Observatory and that what might have been the most significant outcome of all his years of work was abandoned. "The counts were tabulated; I had a lot of ammunition ready to go when I was done with blinking." It was ammunition still frozen in the plate emulsion. The great survey work would be left to Lick and Palomar and the research of Fritz Zwicky and George Abell.

During Clyde Tombaugh's stay at Lowell a total of 3,969 asteroid images were marked on 338 pairs of 13-inch telescope plates and 24 pairs of 5-inch telescope plates, all taken during the trans-Saturnian planet search. Possibly 775 of these asteroids had not been observed previously. Two comets and one nova were discovered, although none was reported at the time. Five new open star clusters were revealed, along with several clusters of galaxies, one supercluster of galaxies, one globular star cluster, and one trans-Neptunian planet. "I wanted to find out for myself what was out there." Although Tombaugh had expected to discover more planets, it was not a disappointment to him that he did not. "A lot of my happy years were there, and some of them were not so happy. But I certainly enjoyed observing, and my heart was in it up to my ears. But I did get weary of the observing in January and February; it got so cold—I suffered in the cold. I sat there, did not have time for exercise, could not leave the telescope."

The telescope Tombaugh was finally about to leave stayed at Lowell, its long career not nearly over. Years later Director Arthur Hoag claimed that with its many projects, the 13-inch had become one of the world's most productive telescopes. If it had consciousness, it might have wondered what would happen without the man it had shared a life with, a man who had never quite given up searching one more pair of plates for hidden treasure, a man who called himself a traveler "going over the next hill, with an eternal hope."

Chapter 9

WHITE SANDS YEARS

A USTIN VICK was looking for a job. He had just graduated in civil engineering from the New Mexico College of Agriculture and Mechanic Arts (later New Mexico State), but in early August 1950 jobs were scarce. He found one in Santa Fe with the New Mexico State Highway Department; on his way north with all his belongings, he stopped for gasoline and started chatting with the attendant.

"No jobs in Cruces for civil engineers?" The gasoline station attendant looked puzzled. "I hear White Sands is hiring engineers." The attendant even pointed Vick toward the house of Roy Dean, the civilian personnel director. Dean was outside mowing grass. Two days later Vick started at White Sands, working on the Mitchell phototheodolites under the direction of Clyde Tombaugh.[1]

That was Vick's second surprise — here at White Sands he was working for an *astronomer*: what was an astronomer doing at such a place? "Nobody had to tell you that he was an astronomer. Just by his attitude and conversation you could tell that he was an astronomer." He instructed his men to "treat [the missile] as if it were a star image. . . . In between launchings, when we didn't have a lot of work to do, Tombaugh would give us astronomy lessons."[2] Officially, Tombaugh was an optical physicist and astronomer; later he was an ordnance engineer. His rank was equivalent to lieutenant colonel, and he was in charge of some seventy-five people. Was this the same Clyde Tombaugh who not five years earlier was at Lowell Observatory, working with a very small group of people in a search for planets? Like Vick, Tombaugh also had taken an unexpected turn in the road.

After V. M. Slipher's conversation, the thought of leaving Lowell Observatory was both scary and invigorating, and despite the hurt feelings, Slipher's suggestion about going into teaching appeared a good idea. Tombaugh had enjoyed teaching physics, navigation, and astronomy. Also, the lecturing he had done as part of his civil defense work had expanded his abilities in speaking before diverse audiences. It really looked as though he would travel in that direction in the coming years.

With his reputation it was not difficult to find a teaching position. While at UCLA he had actually applied and had been accepted for a position at the University of New Mexico. A real problem with the UNM position was that finding housing was difficult, and Clyde thought that for the first year Patsy and their two children—Annette about six, Alden less than two—might have to stay in Flagstaff. Tombaugh's plan: a few more months of blinking and then a move to Albuquerque for a career in teaching with Lincoln La Paz, a well-known meteoriticist who was also a mathematics professor.

Just as the arrangement was completed, Patsy's brother, James Edson, whose rich career would include the first photograph of Venus's atmospheric ring, dropped by to visit—in an official military vehicle. This call was to be more than a family chat; James had arrived with a surprising proposal. "You've got to come to White Sands," Tombaugh recalls his brother-in-law's words. "We have a tracking telescope with a long focal length, and you're just the fellow for it. No one else understands it as you do."

At the White Sands Proving Grounds (WSPG) northeast of Las Cruces, New Mexico, the army was in the process of developing a launching facility for the V2 rockets it had captured from Germany at the end of the war. WSPG had begun as an outpost annex of the army's Ballistic Research Lab at Aberdeen Proving Ground in Maryland and became White Sands Missile Range in 1958. It looked like an awful place to live and work.

With people like James Edson helping to set it up, however, White Sands was far more interesting than it appeared. It was the home of a cache of captured V2 rockets and a workplace for a large group of German scientists, led by Werner von Braun, who had been surreptitiously moved to the United States through Operation Paperclip. At first the scientists had been quartered at Fort Bliss in Texas, not far from White Sands. After the State Department found out about their existence, frantic negotiations ensued and the scientists were boarded into buses, taken to Juárez, Mexico, and returned to become official United States immigrants—from Germany via El Paso.

The V2s themselves were stored in salt mines by Major Herb Karsch, who later became technical director at White Sands. The early V2s had been tested on the shore of the Baltic Sea, at a place called Peenemünde, not far from the Polish border. Those rockets not good enough for launching were dismembered for parts; well over fifty V2s were eventually fired. The nose cone, or ogive, contained all the tools—cameras, spectrographs, and X-ray equipment—to study the upper atmosphere. The balloons used earlier had been able to rise no more than twenty miles; the powerful V2s were now rising more than one hundred miles, loaded with upper-atmosphere experiments. This project was really the dawn of America's space age, and its first harvest of learning was Earth's upper atmosphere. These war rockets provided what Tombaugh called "a marvelous laboratory."

As Edson explained, the scientists—American and German—were gathering at White Sands to launch these missiles and study their behavior in flight. To do that, they needed high-quality optical telescopes, procedures in operating them, and most important of all, a teacher to instruct and motivate the operators to get the most precise results possible. The task really was very similar to photographing the planets with a long-focus telescope.

James Edson had already brought the first tracking telescope down to White Sands while working on his Ph.D. dissertation at Johns Hopkins. The instrument consisted of a 5-inch and a 4-inch, separated by about four feet, together called "Little Bright Eyes." "Tombaugh's talents for optical instrumentation and for precise observation were needed at White Sands," Edson says. "I made the contact between Tombaugh and the White Sands command, and nature took its course."[3]

Nature may indeed have taken its course, but Tombaugh fought it at first. His brother-in-law's arguments notwithstanding, Clyde was skeptical and genuinely worried that all the experience he had had with planetary astronomy would "go out the window." At the time, rocketry was such a new field—White Sands was essentially in its first year—that it might not have much of a future. The only reason he took the position seriously at all was that it involved optics and telescopes, although "the rockets sounded interesting, too." Patsy and James then suggested that he at least go down and see what the place looked like. Thus, Tombaugh went to White Sands to see a V2 launch and its fifteen-minute flight. "We kidnapped him!" Patsy says a bit euphemistically, "We kidnapped Clyde and put him in the car!"

As soon as he arrived, he understood the enormous potential of the place and the great challenge of applying his experience from tele-scope making, dating back to his childhood, as well as his more recent planetary photography, to the tracking of rockets in flight. There was nothing trivial about this assignment. "If it's telescopes," Tombaugh realized, "I'm there."

During Clyde's trip Patsy stayed in Flagstaff, uncertain whether her husband would be interested in her brother's plan. Then the phone rang one night: "I remember him calling and saying, 'Guess what they're going to pay me?' He was just astonished!" The salary offer was $4,900 in 1946, and it went up soon afterward.

The ultimate irony of moving to White Sands was that, from a philo-sophical viewpoint, it did not involve as much of a change in career as teaching in Albuquerque would have. Here Tombaugh could con-tinue working with optics, applying many of the same principles he had learned while working at Lowell's large refractor or building a 9-inch reflector in Kansas. "The problem," as Tombaugh put it, "was actually an application of planetary long-focus photography to missiles in flight."

Tombaugh was probably the first rocket scientist coming from as-tronomy and thus perhaps the first astronomer of Willy Ley's "third era" of astronomy. In the first, the astronomer looked at the stars with unaided eye; in the second, the telescope was invented, and in the third, people visit the planets with spacecraft.[4] Other astronomers came later, including Dirk Reuyl and Henry Cobb, and still others, particularly from Harvard and Yale, consulted for the program. Astronomical work with the tracking telescopes did have one direct function. Calibrating the readings on all the telescopes was done using the Moon, a process conveniently aided by the fact that in the years around 1950 the Moon's orbit brought it to within four degrees of the zenith at the latitude of White Sands.[5]

The First Winter, 1946–47

When Tombaugh passed through the main gate at White Sands Proving Grounds, he entered a frontier army base and a life vastly different from any he had known before. His youth had been spent on a farm, in the company of just a few people. His Lowell years had been spent on a huge mountain site occupied monastically by a handful of people. Except for his teaching, Tombaugh's life, though challenging and busy, had been sheltered. Now he was something utterly different, a civilian in the military, with eventual responsibility for many people.

Although the military people and the civilians had come from different backgrounds, there was surprisingly little friction between the two groups. Military people had to salute officers; civilians did not: "we thought of ourselves as a team. There was not the type of difficulty we had up at Flagstaff at all." The people Tombaugh encountered were very bright—one group, known as the First Guided Missile Battalion, consisted of some of the army's most brilliant people.

Living in the army barracks, eating in the mess hall with the GIs and everyone else, he found conditions unusual but tolerable. Of course, the advent of air conditioning there was still some time away. For the first year Tombaugh did not have an office at the base: he only half-joked that his office was in an army jeep out on the range somewhere. "Clyde immediately 'locked on' to the problems and opportunities at White Sands," James Edson writes. "He began within a few days of his arrival to produce important results. Probably he first felt satisfaction with accomplishment rather than enthusiasm."[6] His first major contribution came late in 1946. When Tombaugh arrived, the favored launch time was 11 A.M., a time of day when "seeing"—an effect involving the steadiness of the atmosphere—was at its worst. His pictures were criticized for being out of focus by those who could not recognize bad seeing, and Tombaugh's repeated suggestions for launch times early or late in the day went unheeded. "Turbulence? What's that?" Finally, early in December a V2 countdown was delayed until late in the afternoon. When the launch finally took place, the Sun was behind the cameras and low in the sky. Tombaugh's team got sharp pictures of the V2 up to 110 miles. "We could see the shape of the missile!"

It was an accident that the December 5, 1946, flight left so late. "Henry Cobb was down from Aberdeen at that time. That evening they developed the film and got prints from fifteen or twenty different frame regions. They were phenomenal! They had never gotten anything like that with such detail." At the firing evaluation meeting the following day the pictures "stole the show. The tracking telescope had been considered kind of a joke and even ridiculed. From then on, it was respected." That day in December marked Tombaugh's coming of age at White Sands; he had used his optical skills to accomplish something unique.

The improvement in tracking was so great that from then on Tombaugh's suggested times for launches, early morning or late afternoon, were taken seriously and formulated into launch windows. The pictures were taken back to Aberdeen Proving Ground, where officials were amazed that the small 5-inch camera, designed by James Edson, had produced such fine results. "Aberdeen went hog wild over that. From then on, the emphasis was on more tracking telescopes. We had it made!"

Tombaugh's second idea involved the paint pattern on the V2s. The original paint design, an unimaginative combination of black and yellowish-green, was not visible over long distances. The Germans had relied on their telemetry and their measurements of the Doppler effect of the missile's radio waves to learn what they could about their launches. Tombaugh proposed a pattern of flat black and flat white, and when that suggestion was adopted, the optical tracking became far more effective. The paint pattern on the fins alternated white and black so that rates of the missile's roll could be determined accurately. Additional diagonal stripes acted as a measure of the resolution of the tracking telescope film. That was not necessary unless only one camera recorded the shot; afterward someone evaluating the film could compare the rocket's foreshortened image with its actual diameter.

Meanwhile, Patsy was trying to set up family life in Las Cruces, which in 1946 was hardly less primitive than White Sands. When she and the children arrived in October, the hot and windy afternoon was a depressing beginning to their new life in New Mexico. The streets seemed to be seas of trash that blew onto front lawns. Since Clyde was still living on site, he rented a room at the El Molino Motel for Patsy and the two children. Built in Spanish style, the hotel was an interesting place in a nice wooded area. But no sooner were they settled than Tombaugh said he would be leaving right away for a two-week trip to Aberdeen Proving Ground in Maryland. That first airplane trip in October 1946 was something he would long remember. The aircraft had no seats; the passengers had to stand between pilot and copilot for the eight- or nine-hour trip. Tombaugh got airsick and had to lie on the floor to hold himself together.

Several times Tombaugh flew to Aberdeen on an aging C47 whose creaks and groans did not inspire much confidence; he got to know the flight crew well. His worries about the plane were justified when he learned that flying over Memphis, the plane had fallen apart, killing everyone. "I rode seventeen thousand miles on that plane. I was absolutely petrified. These guys had all been killed, and it really got me."

After about three months in the motel, Patsy decided that the family should buy a home. Accordingly, Tombaugh inquired about his future at White Sands and was happy and surprised to learn that he would most likely have a position for at least five years. "In those early days," Tombaugh explains, "people were not sure the rockets would be here to stay. Only a few missionaries could really see that they would. There was a lot of skepticism in a lot of places." The rockets' military value was uncertain, and possible scientific benefits were not given much thought. Many Las Cruces residents had little use for the influx of rockets and missiles.

With this confidence, they began to look seriously for a house and finally found one at 636 South Alameda for $10,200. The early months in that house were miserable. Patsy got the flu; Tombaugh was working virtually around the clock. The house's old adobe-chinked roof leaked so badly after rains that once the water shorted out the wiring. Putting raincoats on the bed and newspaper on the floor offered the only protection. A new roof, they learned, would cost several hundred dollars but could not be installed until the rain stopped long enough for the old one to dry out. They also had to shut off the gas.

With drier spring weather, life rapidly improved. A new roof was added to the house, and they ended up living there happily for twenty years. In the meantime, Tombaugh was having the time of his life at White Sands. In some respects White Sands was a lot more relaxed than Lowell had been. The frontier atmosphere of the place made it very hard work but fun. "It was a relief to go from having been, in a sense, snubbed, to holding a respected position."

The Tracking System

Tombaugh's first major job was to pick sites for the tracking telescopes. In the first few months the team members had learned much about tracking, including losing a missile after fifteen miles because the sunlight was too bright to let them see it. A jeep was assigned to Tombaugh; "I must have driven two thousand miles over that range looking for sites." There were few roads on the brand new range, so he just drove across the sands. One incline was so treacherous that the passenger had to hang outside to keep the jeep from overturning! One site Tombaugh chose on San Andros Peak was somewhat difficult to reach; the army installed a cable and bucket that climbed some three thousand feet over a mile to reach it. During fine weather the arrangement worked well, but the site was prone to lightning during storms, and one of the men was almost struck. The cable started to have trouble, so eventually the site was abandoned. "But it gave such a superb view!" The most distant site Tombaugh chose was on Mule Peak in the Sacramento Mountains, two hundred miles across formidable terrain from the base. To man that station, a crew would arrive Monday and stay through Friday.

Essentially, there were four types of optical tracking. The 10-inch tracking telescopes, and later the more powerful 16-inch Intercept Ground Optical Recorders (IGORs), were part of the Attitude and Events Section. The cinetheodolites, mostly made by the German firm Askania, captured from Germany after the war, were designed to track the missiles for trajectory purposes. Three Bowen-Knapp fixed-camera stations, one mile west, south, and east of the launch site, respectively, and using 5-inch film at thirty frames per second, predated Tombaugh's arrival. Finally, the fixed Princeton ballistic cameras (also known as star cameras) compared missile tracks against star fields for the first few thousand feet of launch.

To set up the tracking telescopes, the army simply collected some discarded M45 fifty-caliber machine-gun mounts and brought them to White Sands, where the guns were removed and the telescopes added. The telescopes were in welded steel tubes. Tombaugh proposed the use of Newtonian reflecting telescopes as the answer to the need for large-aperture telescopes that were free from chromatic aberration. These reflectors had one responsibility: to take a picture with sufficient scale that details on the rocket were visible over large distances. The telescopes were numbered T1 ("Little Bright Eyes") through T5. T2 was closest to base and sometimes used a diffraction grating for spectra of the flame that was recorded off to the film's side; T5 had the mount with the worst backlash. Tombaugh tried to take care of that problem with counterweights of lead-filled cans, but they increased the friction so that the telescope was very hard to move.

The instruments were set up in two lines, one on the east side of the launch site for morning firings and the other on the west side for afternoon firings. Because the missiles moved so quickly, one person had a rough time handling both azimuth and elevation motions at once. Two people were used for tracking. Using 2-inch, 10-power cinetheodolites, one moved the azimuth controls, the other elevation. The system depended on the ability of each tracker to keep his motion accurately following the object. In the meantime, the telescope was recording the launch—or an embarrassingly blank sky—at sixteen frames per second.

In order to get high-dispersion spectra of the materials coming out of each rocket, the T2 camera, situated closest to the launch site so that it could record the strongest intensity of light, occasionally covered launches spectrographically. This program had its genesis in James Edson's dissertation on spectra of V2 fuel combustion. It was necessary to be close to the trajectory to get these images, and thus at times Tombaugh and Edson were three hundred yards from the pad. They observed two launches this way. "We put cotton in our ears, and it was an experience I'll never forget. The vibration just shook us like mice. An accident, an explosion at the launcher, would have killed us. After two launches we figured we had used up our luck and we stopped."

After the Edson experiment Tombaugh saw some value in attaching a spectrograph to the T2 telescope. The purpose of having these emission spectra was to analyze the fuel output. Typically, there was only one strong yellow D-line, indicating sodium in the fuel. After one failure the spectrum showed several hydrocarbon lines, indicating only partially oxidized alcohol. From the spectra Tombaugh concluded that the oxygen pipe had somehow become clogged, preventing complete burning.

As the launching program expanded and other types of missiles were added to the arsenal, more effective tracking telescopes were needed.

The Nike missile was accurate enough to intercept targets in midflight. Developed specifically to see far enough to record these Nike missile intercepts, four IGORs were built at Aberdeen Proving Ground's U.S. Army Ballistic Research Lab. A related system, called the Intercept Target Optical Recorder (ITOR), was attached to the drone for the same purpose. Obviously, the ITOR data was available only if the drone survived the intercept.

Partially developed by Tombaugh, the IGOR telescope used a 16-inch f/6 mirror. The IGORs were Newtonian reflectors in which light was reflected from a primary mirror to a small secondary mirror, then through a hole in the side of the telescope tube; the camera was mounted on the outside of the tube. Thanks to the idea to use Barlow lenses, originated by one of the young telescope men, Joe Marlin, the power of these instruments was increased dramatically. He and Tombaugh assembled one; its field of view had some distortion so could not be used for precise attitude measurements, but its recording of special events such as intercepts was flawless. An early success with it involved a Nike missile launched to intercept an old aircraft controlled by radio. The pictures the telescope produced clearly showed the Nike hitting the plane's fuselage.

The early telescopes were set on trailers with only tarps — actually Conestoga wagon covers — to cover them. At the beginning of an operation the covers would be rolled off, the telescope would be cranked to its correct position, and work could begin. Tombaugh would clean the telescope mirrors himself with distilled water and cotton. In one carefully rehearsed caper, someone casually alluded to having cleaned one of the IGOR telescopes. "How did you do it?" Tombaugh asked anxiously. "Oh," the man drawled, "we used a scouring pad and steel wool and did a real good job!"[7]

The theodolites measured and recorded azimuth and elevation angles on photographic film. If the exposures were taken at random, the instruments were called *phototheodolites*; for continuous exposures pulsed at a set rate, the instruments were *cinetheodolites*. Although the images on the early theodolites were not detailed enough to study rocket attitude or any specific events, they were adequate for calculating missile trajectories if good records from at least two of the instruments were obtained. For its time, however, the German Askania design was quite sophisticated. On the corner of each cinetheodolite frame the azimuth and elevation were recorded to a minute of arc (about one-thirtieth the moon's apparent diameter).

As a rocket passed through an elevation of thirty miles, its image got so small and indistinct that the cinetheodolite did not track it properly. Tombaugh developed an idea to increase the telescopes' powers by replacing their 4-inch lenses with 6-inch mirrors with cassegrain foci of seven, ten, and fourteen feet. There was not much support for this improvement from White Sands until he demonstrated how well the new system would work; nor was there time to go through the army's normal requisition process, so Tombaugh's men manufactured the mirrors themselves. As a result, the instruments could see the shape of the cone from forty miles.

It was up to the electrical measurements branch, headed by Frank Hemingway, to provide synchronization signals for the cinetheodolites and equipment for each station to record the times. The cinetheodolites could film a launch at one, two, or four frames per second. Someone reducing the data on azimuth and altitude could interpolate to derive a position to an accuracy of 1/10 arc minute.

The station that sent the pulses to expose the cameras was set up some three miles west of the launch pad. The electrical measurements branch arranged for the cameras to fire at precisely the same time, taking into account even the velocity of light between stations closer and farther from the launcher. Thus all photographs were coincident to within one millisecond.

Since neither the cinetheodolites nor the long-focus tracking cameras were good for tracking right at the launcher, a different type of instrument, the ballistic camera, was developed by Henry Norris Russell, the famous Princeton stellar astronomer who consulted for Aberdeen. The ballistic camera got its name because it recorded the ballistic path of a missile; it was fixed but had a tracking screen that would block all areas of the plate except a small strip near the missile. After the firing the camera would be left exactly in place. That night, the camera would expose the same plate on the field of stars. When the two exposures were compared, very precise determinations of azimuth and elevation could be made. During most launches three ballistic cameras were operating. The night sky could be photographed in a short exposure any time that night, as long as the time was noted precisely.

The details of the ballistic camera operation were developed by Tombaugh and were headed after 1951 by Jed Durrenburger. A successful record involved at least two fixed cameras whose positions had been accurately determined. After the powered flight phase, tracking of a night-launched missile could continue if it had a strobe attached to it; thus each strobe would record on the 8-by-10-inch glass plate and 103-F (or 1N for red) film emulsions as a star image. The star fields would later be exposed on the same plate. During night launches a calibration star exposure would be taken both about thirty minutes before launch and some fifteen minutes after the flight. The ballistic cameras were particularly effective; on one especially busy night the cameras recorded twenty-seven launches.

The Organization

Other than Tombaugh's optical measurements branch, there were several branches working with the tracking project. Hemingway's electrical measurements branch provided time recording and related support. A third area was Ben Billups's ballistic data reduction, and there also was a radar branch. The actual firing decisions were made by Pappy White, from General Electric. Like all branches, Tombaugh's division had its own military officer to take care of some administrative duties, including the ordering of any materials that were needed. It was a task Tombaugh thankfully relinquished, since he did not want to spend time on administrative formalities. Essentially, Tombaugh had brought the same attitude to his job at White Sands that he had used in his planet hunt: he wanted to concentrate on the essence of the task at hand. He found that easier to do at White Sands than he had at Lowell.

The job included a great deal of teaching, for his men had to be trained in their jobs. The students included some foreign military people, even on one occasion France's minister of defense. To these people Tombaugh's astronomical work, including his discovery of Pluto, was apparently irrelevant; he was an ordnance engineer in charge of obtaining ballistic information.

The optical measurements and electrical measurements branches lasted until 1952, with Aberdeen Proving Ground having responsibility for research and development and White Sands with operations. A major reorganization changed the responsibilities of institutions and people, and the air force ceded its R and D operation to the army. Frank Hemingway then took over the optics branch, and Tombaugh was transferred to R and D. With the increase in organization, Tombaugh was not as happy as he was during the more freewheeling early days, and it was about that time that he began planning his proposal for a search for small natural satellites (see Chapter 10).

White Sands Life

Getting to and from work was a large part of a White Sands day, since the twenty-six-mile drive each way at the time lasted close to an hour. In June 1947 Hemingway and Tombaugh started car pooling together. Hemingway remembers one aspect of these drives:

Most of us carried work home, of course, and worked odd hours — all night sometimes — and we carried briefcases. But Tombaugh's got bigger, and bigger, and bigger — he had one of those old folding expandable ones — finally got it out to where you could swear the leather was going to break. He could just barely get it in, and the car-pool members grumbled about the size of it. We all wondered what he was going to do, because we knew he couldn't stuff any more in it. Well, lo and behold, one day he comes out with two briefcases! Not only carried the big one, he had another one! We threatened to charge him for another passenger's place in the car pool.[8]

Although a number of people developed the methods for mathematical data reduction of the missile journeys, Tombaugh pioneered the process by which detailed photographs are taken of a missile in flight. To keep up with changes for each new rocket, Tombaugh would meet with the engineers to learn precisely what kind of tracking data was needed. If, for example, the initial velocity was very fast, he would have to use the instruments farther away from the launch site. He also decided how best to filter the exposures to reduce fogging of film by excess light.

During one test a missile fired for only four seconds. Tombaugh was operating a camera right under its trajectory: "the shock wave was so intense it knocked me right to the ground on my knees!" On days of good weather Tombaugh would witness a launch from one of his stations. It was only on the marginal days, with clouds passing by, that he would be at the blockhouse. When the cloud conditions were right for at least three of the stations to track the launch, he would authorize it and the missile would launch soon after. The timings on films started at the instant of launch.

On partly cloudy days Tombaugh would join the launch crew in the blockhouse and watch the sky carefully. At certain configurations of cloud it was possible to launch a mission with sufficient cinetheodolites to get the trajectory recorded accurately. Once they tried to launch with an 80 percent overcast. Using a map and information about cloud layers from the weather station, Tombaugh identified the stations that would be under clear sky during the rocket's powered flight. On that day he found that only two of the eight cinetheodolites could follow the flight through unclouded sky. If one of the films should jam, Tombaugh protested, there would be insufficient data to calculate the trajectory. His fear was justified—the mission was launched, and one of the cameras jammed.

Informally, under Tombaugh's leadership these tracking telescopes were used for both visual and photographic observations of the Moon and planets. Because of the machine-gun turret mounts, exposures had to be very short. The closest of the 10-inch telescopes to the base was T2, and it probably was used for this purpose more than the others. Although the telescopes were never intended for these observations, one occasionally might need "test objects" for alignment or for pointing practice, so why not have some fun? For visual looks, the camera would be pulled off and replaced with an eyepiece.

Inconsistent focus was one of the early problems the crew faced. To isolate the difficulty, Tombaugh arranged for certain stars to be photographed in broad daylight at different focal lengths and different elevations. He taught the men to calculate the altitude and elevation of a star at a specific time of day. "I could run out there in the daytime," Joe Gold says, "look at my watch, and compute the altitude and elevation of a star like Vega, and lo and behold it would appear!"[9] The program, conducted whenever there was time, lasted for months and determined that the cause of the poor images was poor seeing. It was one of Tombaugh's more significant finds. Even the shorter-focus cinetheodolites suffered from poor seeing. The solution, Tombaugh proposed, was to raise their mounts so that they would be high enough off the ground to avoid the effects of near-ground turbulence. A compromise was reached, since with too much height the cameras would lose stability.[10]

As White Sands expanded, other rockets arrived for testing. The Honest John was rather innovative in at least two respects. It was the first large U.S. booster to use solid fuel, and it was named after a Juárez, Mexico, liquor dealer. As one version of the story goes, one night the group that brought the missile to White Sands went to Juárez to buy liquor from a man who called himself Honest John and claimed that his liquor was the best. When the men returned, they gave it traditional scientific tests, found it to be the worst they had ever tasted, and named their missile Honest John.[11]

The night firings were particularly interesting. In December 1946 a firing was scheduled to release artificial meteors to study the ionosphere. The single-stage booster traveled 114 miles before starting down and leaving a beautiful reddish trail. The predawn launches — perhaps six a year — had special problems and a special beauty. They were trackable during their powered flight phase; afterward, when the missile got high enough, it would appear again, this time lit by the Sun. Late night and predawn firings were the roughest on Tombaugh's schedule. Whenever these firings were postponed, Tombaugh would have to inform his full team, scattered everywhere. During the first years V2 launches took place every two weeks. As more and more different types of rockets were launched, firings became more frequent; in Tombaugh's later years they occurred almost every day, sometimes two or three times a day.

Tombaugh's love for animals was legendary at White Sands. Although the fauna generally were tame, before one early morning launch the team found a rattlesnake comfortably coiled directly under the camera. Tombaugh would criticize his men for throwing rocks at rabbits to force them away from the cameras. One day, however, a rabbit did come by, then backed away over a small mound. The others were trying to hit it, so Tombaugh threw a small stone himself just to scare it away. "I just gave it a little toss, you know, hoping it would scare him and he would run away. I heard a clunk. It had landed on his head, and he was deader than a doornail. I felt so sick about that! . . . I was trying to protect him, but then I killed the poor little animal." As others found out about this accident, he was teased relentlessly.

Learning

Informally, there were two kinds of trackers for these instruments, the "apprentice" and the "good." The apprentices literally learned under fire, during actual launches, and had to become good quickly. Either kind of tracker could make errors, and during many launches the number of different kinds of errors that could be made was legion. Forgetting to remove the lens cap, not loading the film properly, not turning on the timing, and not having the camera pointed at the launch pad at firing time were among the possible errors. "Clyde's reaction to these errors," says Joe Gold, "was far more benevolent than that of the military officers. That's because he had been there, he understood it . . . this tended to create a great loyalty within the work force toward Tombaugh." [12]

In some respects army procedure was not quite up to launch demands, especially because the tracking stations needed to be fully manned. Although Tombaugh, of course, was not subject to army discipline, the enlisted men under him were. Tombaugh remembers once before a firing that a team member was missing from a cinetheodolite station because he was on KP. "They should have told me; how was I to man that station?" Standing inspections were also an inconvenience during firings. Once when a launch delay meant that the crew missed dinner, the kitchen would not feed the men. Tombaugh walked into the kitchen, saw the cook making doughnuts, and simply started to gather them up for the crew. "The cook was so angry I thought he would jump over the counter and grab me by the throat!"

Losing the target in the camera happened from time to time, although only once was the reason unearthly. On one flight a rocket happened to pass very close to another object visible in daylight and about as bright as the rocket at that instant. Confusing the two, the trackers followed the wrong thing. When they realized that their "missile" wasn't moving, they suspected something was wrong and said so over the command line. "Well, folks," Tombaugh announced, "what you're looking at is planet Venus!"

When Tombaugh was away at Aberdeen, one of the V2s, a victim of incorrect wiring, went about 180 degrees off its course. Instead of traveling north uprange, it headed south directly toward heavily populated El Paso, Texas, and neighboring Juárez in Mexico. The German scientists in the blockhouse saw the problem first, but their panicked shouts in German were not understood by the firing crew. The miscreant V2 traveled farther and farther south and crossed over the border. A celebration was going in Juárez when the V2 dove harmlessly into a cemetery, its thunderous crash somehow blending in conveniently with the fiesta. Launches from White Sands were stopped for about two months after that event while the army learned why, as Tombaugh put it, "we bombarded a foreign country with a missile!" The event intensified a debate about the importance of having White Sands develop so close to Las Cruces. The argument raged on in the newspapers between those who worried about the safety of Las Cruces with out-of-control missiles arcing about overhead and those who were in favor of the development of White Sands as a major base and the consequent growth of the town.

A common early problem after powered flight involved the oscillation of telemetry signals. The Germans had attributed it to the rocket's hitting the upper atmosphere and suddenly arcing downward. Better pictures forced a new interpretation: an instant before a rocket stopped firing at a height of about twenty miles, the last bit of fuel fired not symmetrically but instead toward one side, causing the rocket to gyrate and the signals to oscillate. These oscillations did not occur at every launch, but whenever they did, experimental data on any upper-atmosphere research were hampered. Thanks to this discovery, a solution was devised: primacord around the fuel pipes would be blown a fraction of a second before the fuel was exhausted. This way the fuel shutoff was clean and symmetrical for every flight.

A third early success for the tracking telescope was detection of the source of explosions long after launch. If an explosion occurred on an early V2 near its apogee of more than one hundred miles, it was difficult to tell from which part of the missile the explosion came. The long focus of these telescopes produced pictures good enough that the exact site of an explosion could be found. These three developments all took place within the first year and all were a direct result of the work produced by the tracking telescopes.

Tombaugh saw more than fifty V2 launches from White Sands, including one particularly balky "hangar queen" that returned five times before it finally achieved a launch. Another missile giving problems on its launch pad caused delay after delay, and Tombaugh broke in on the command line: "If that missile won't do its job, fire it!" The rockets were forty-six feet long and about sixty-two inches in diameter. Without fuel the systems weighed four tons; at launch this weight was increased by about fifteen tons of fuel. (If the rocket couldn't be launched in one day, the dangerous activity of removing all the fuel would take place.) The Germans had designed the V2s to carry a metric ton of explosives over a distance of two hundred miles. Wherever they hit the ground, they left impact craters twenty feet deep and more than seventy feet wide.

It was not always possible to direct a V2 safely, and a missile off course because of a guidance-system failure is a scary thing. Once a V2 went straight up, not following its programmed instructions to turn northward. After its burnout it simply turned tail and headed down to a place close to the launch site. Four tons of metal slammed into the planet from a height of one hundred miles. Fortunately, it fell between two of the Bowen-Knapp fixed camera stations, leaving its crater less than a hundred yards from one of them. Although no one was hurt, the men were badly frightened. Tombaugh recalls: "I was in the station south [of the reentry path] and saw the thing coming down, and I was just petrified. It was exactly in line with that station. Then this cloud of dust came up, and I thought it had got those guys."

After that close call Tombaugh ordered his team to dig slit trenches and not to wait until they knew where the missile, cascading in at more than four thousand feet per second, would land. As soon as they saw it approaching, they were to rush to the trenches. With this precaution one could probably survive as close as forty yards from a missile's impact point. The other rule was "Scatter!" On one occasion Patsy was watching a V2 launch that did not go properly toward the north. Patsy's group was out in the open when it was ordered to scatter; in that way there was a chance that one person would be hit but far less chance probability that the entire group would be struck.

One of the finest examples of Murphy's Law–type problems occurred during the upper-atmosphere research. An experiment by Marcus O'Day in 1947 planned to use a large transmitter to trigger the ionosphere to create an artificial aurora at the apogee of a V2 flight. The flight had to be postponed, because—as far south as White Sands—there was a real aurora that night![13] Aurorae may occur once or twice per eleven-year solar-activity cycle as far south as White Sands. An experiment involving a camera to photograph the Earth was more successful. Designed by Clyde Holliday, the camera survived to provide pictures of a curved planet from the then-record height of one hundred miles.[14]

One of Tombaugh's scariest moments came during a launch in which the rocket had been set for horizontal flight. Whizzing northward, it went several miles before a carbon vane rudder was damaged; then it started heading back at the men. The horizontal flight was so low that radio contact with it was lost, so that it couldn't be shut off. "We were right in line with one of its gyrations. I told the fellow, 'We'd better get down behind this boondock and just hope it lands before it gets to us, or lands beyond us.' For about five minutes we didn't know whether we were going to live through it or not. It dove into the ground about a half mile from us and then exploded."

One of Tombaugh's last major missile projects at White Sands involved the design of the small missile range. The small missiles, mostly antitank weapons, required their own cameras for tracking. These tests used objects moving so rapidly, always near ground level, that cameras needed to be set up right along their tracks, but without their operators. Working with Frank Hemingway and others, Tombaugh designed a remotely operated system to record these launches. He was involved in every aspect of the design, including where to put the range (they didn't want these missiles darting across nearby Highway 70). Automating the tracking cameras was a good idea. The missiles are supposed to travel close to ground level, but still a few feet above the tracking cameras. During one firing, a missile went too low and slammed right into one of the cameras. The camera shattered, but the lens, after flying off on its own, survived intact!

The Bumper

One of the most exciting things tried with V2s at White Sands was a second-stage WAC-Corporal addition called the Bumper. Racing skyward to a height of two hundred sixty miles, the rocket painted a vision of the future for Tombaugh, who foresaw an age of orbiting satellites. In 1949 the fourth attempt to launch the Bumper two-stage rocket succeeded. The rocket lurched upward some two hundred fifty miles over White Sands, and the IGOR tracked it for the first fifty. It was the first time that powered flight had been seen so high. "The path was so high that the jet spun out like a comet's tail," Tombaugh recalls; "it was the most magnificent thing you ever saw." The sight of the booster arcing into the sky was a thrill to the men tracking its graceful flight, as well as to the visionary who hoped to see the program go further.

This was a time of dreams, and Werner von Braun and Clyde Tombaugh got along well because both had vision. After the success of the Bumper rocket, von Braun and Tombaugh considered the possibility of a third stage that would impart enough velocity to put a satellite into orbit. Their discussions culminated in a special meeting in Washington. With other scientists, Tombaugh attended; it would be his role to develop the optical instrumentation. Von Braun's satellite proposal was called Project Orbiter.[15] This new project appeared the obvious next step, until the group was ordered specifically to end all discussion: there would be no third-stage rocket. Tombaugh recalls the attitude at the Pentagon at the time, with people who did not see the value of the V2s in the first place. "There are harder fights fought in the Pentagon than are fought on the battlefield!"

Of course, that view changed suddenly on October 4, 1957, when the Soviet Union launched Sputnik I. National policy turned 180 degrees overnight, and von Braun was suddenly asked how soon he could launch a satellite. The answer was ninety days. Explorer I was launched, Tombaugh recalls, eighty-nine days later.

Chapter 10

ASTRONOMY AT WHITE SANDS

ATSY MARRIED a very serious person in 1934. It was only in Clyde's later years at Lowell that he began to develop the sense of humor for which he is now as justly famous as he is for discovering a planet. While a graduate student, he met a good punster who introduced him to this beautiful aspect of the English language. By his White Sands' years, Tombaugh's puns had become an important part of his personality, and also served an important technical purpose, as Jed Durrenberger remembers: "They helped to keep a meeting on a stable keel and keep things light-hearted."

At White Sands his puns were even rated; Tombaugh would often laugh at his own jokes with a sharp "ha!" as he straightened up in his chair. A pretty good pun would get two or three "ha's" or "ho's," and the best would get three or four and then he would slap his knee. Examples include riding a bicycle that's "too tired" (three-ho) or:

Joe Gold: "Clyde, I built this house. I did everything."
Tombaugh: "You laid the bricks and everything?"
"No, Clyde, I hired bricklayers."
"Oh, I'm mortarfied." (four-ho kneeslapper)

Mars

Tombaugh's initial fears were right about losing observing time at White Sands. During the first few years there was little opportunity for any recreational planetary observing. Tombaugh was enjoying his new

147

work, relishing it, but he was "very hurt" to be missing the chance to work on planets. By 1950 the infrastructure had been set up for optical tracking at White Sands and Tombaugh's tasks had become more routine, so some time did become available for observing. He asked his father to send the 9-inch from Kansas which had started Tombaugh on his road to Pluto. Also, he mounted his two 12-inch reflectors together in his backyard.

Of all Tombaugh's Mars sessions, one at McDonald Observatory's 82-inch reflector in March 1950 was possibly the best. Gerard Kuiper, a top planetary scientist who later founded the Lunar and Planetary Laboratory in Tucson, had invited him to come for a two-week observing run. Thanks to a very strong west wind ("I thought it was going to blow the dome off the mountain"), the seeing was superb. At the time, the 82-inch mirror had an unfortunate zone of poor quality in the outermost part of the primary. Tombaugh also noticed that the aperture was too large and suggested that they diaphragm the instrument to 27 inches. Kuiper wanted to use the full aperture of 82 inches. "Gerard," Tombaugh argued, "you know that mirror is not of the quality, and neither is the atmosphere; you know that will harm you. The irradiation [of bright areas onto darker ones] will kill you!"

Tombaugh thought that the image would be improved if he could somehow diaphragm between the shadow of the secondary and the edge of the primary. Just covering the outermost edge—"a sliver of the outer part"—made an enormous improvement. Kuiper objected to putting a diaphragm over the main mirror, however, so Tombaugh suggested a compromise of placing a small diaphragm over the secondary. Kuiper didn't want to do that either. Now almost in desperation, Tombaugh suggested a last alternative: why not place a small piece of cardboard on top of the small cassegrain tube that juts out of the center of the large main mirror? That Kuiper agreed to, and he found an eyepiece cover that was the right size. By testing with a small cardboard behind the eyepiece, Tombaugh calculated that a 2-inch aperture cut, yielding an effective aperture of 27 inches for the telescope, would be the most effective.

"Kuiper was just dumbfounded. I could see canals resolved where before I couldn't see anything." The following night they stopped the instrument down: "That's when I saw canals resolved into multitudes of little separate dark patches in a general alignment—not even connected. [Percival] Lowell had seen those as a continuous line. . . . Oh! the seeing was just exquisite. It was razor sharp at 660 power. I've never seen anything like it."

The new decade brought Tombaugh at least peripherally back to astronomy. In 1950 he attended a conference at the University of Michigan on galaxies. In what was possibly the first great debate about the numerical value of Hubble's constant, which defines the distances to galaxies, Hubble had assigned a distance of nine hundred thousand light-years to M31, the great galaxy in Andromeda. Unfortunately, such a distance would mean that the globular star clusters in this galaxy were one and a half magnitudes, or about four times, fainter than those in our own galaxy. Stunned by the galactic chauvinism of this discussion, Tombaugh rose to make this comment: "Gentlemen, I could more readily believe that something is wrong with the period-luminosity function than I could believe that the globulars in other galaxies are systematically fainter than those of our own. . . . You ought to double the distance to M31." Actually, the situation was more complicated. In 1950 the nature of the globular clusters in M31 was in some doubt since they had not yet been resolved into stars (that was still a year or two away), and there was some additional evidence for the 1.5 magnitude discrepancy.[1] The accepted current distance to M31 is about two million light-years, a figure Walter Baade, one of the great cosmologists of the time, had in mind at that same meeting.

Mariner IV's discovery of craters on Mars in 1965 fulfilled a little-known prediction that Tombaugh had made fifteen years earlier. Before the 1950 Michigan meeting, Tombaugh went to the American Astronomical Society meeting in Bloomington, Indiana, to present a paper that predicted that Mars would have impact craters like those on the Moon. These craters could not be seen well, he explained, because Mars always presents a full or gibbous appearance, and the delicate structure of craters requires a very oblique Sun angle for them to be visible easily. Because Mars is relatively close to the asteroid belt, its surface should be strewn with impact sites.[2] Each impact, he thought, would fracture the Martian crust outward like spokes in a wheel, and he reasoned that the central craters would be seen as dark oasis-like features with the spokes as canals. At other times Ernst Öpik and William Pickering had made similar predictions.[3]

Although Tombaugh had likely gone further than anyone since

Lowell into studying the geology of Mars and wrote several papers on the subject, this work remained relatively unknown. Perhaps Pluto was a curse as well as a joy, for the enormous public impact of its discovery cast a shadow on the perception that a few other scientists had of him. People who were convinced that Tombaugh spent his life as a "professional famous person" would not be moved to take his other astronomical studies very seriously. Furthermore, the circumstances of Pluto's discovery concerned some people for decades. The White Sands Pioneer Group, for example, consisting of those who were at the missile range before 1958, started a White Sands Hall of Fame, and naturally Tombaugh was one of the early nominees. When he was being considered, however, one scientist wrote that Tombaugh's role in the discovery of Pluto was "a myth," an accident.

For better or worse, Tombaugh's fame is linked mostly to this discovery, despite his other contributions. Possibly because of the importance attached by the nonscientist to findings of new objects in our solar system, Tombaugh's other notable scientific work has not received much attention.

> I really would have liked to have had a greater reputation in Mars study than appears in the literature. Although I did write papers, everyone thinks the greatest thing I did was to discover Pluto. This was somewhat disappointing because I did things that were fully equal to Pluto. The work out there [at White Sands], the study of Mars, finding the supercluster of galaxies. But all they think of is Pluto. This is a disappointment in that the public did not attach importance to the other things that I did. They didn't seem to understand that. From the standpoint of a real contribution to science, it isn't always the flashy stuff that really counts.

In 1956 an excellent opposition of Mars aided Tombaugh's reemergence into planetary astronomy. Despite the pressure of his work both at White Sands and in a satellite-search program, he observed Mars assiduously. He also set about to finish the 16-inch. Construction resumed in 1957, this time with the assistance of his young son, Alden, who helped wire-brush the rust off the huge steel superstructure.

Although Tombaugh claims that he avoided astronomy in his discussions with his crews, Joe Gold disagrees: "Clyde can't do that. He's constitutionally incapable of it. He was always preaching astronomy.

He would talk about Pluto anytime you asked him; he would talk of his experiences in finding it."[4] Once in a great while it was possible to use the tracking telescopes for other work, such as taking pictures of planets. Anything but the shortest exposure presented a problem, however, because the telescopes were not equatorially mounted.

In addition to giving formal military lectures on optical tracking, during his White Sands years Tombaugh occasionally gave public lectures in town, and also at the base, about astronomical matters. In 1951 Tombaugh, Walter Haas, and a few others founded the Las Cruces Astronomical Society, with Clyde as first president. Meetings were often held in the Tombaugh home. Haas had already developed a national reputation in amateur astronomy through his founding in 1947 of the Association of Lunar and Planetary Observers. The time was ripe for fledgling astronomical societies, for a few years later an eighteen-month period of international cooperation in science called the International Geophysical Year (IGY) started. It spawned an artificial satellite monitoring program that was ideal for amateur astronomers. Founded by Fred Whipple and known as Moonwatch, this Smithsonian Institution program was essentially the best early way of tracking these satellites. It was possibly the first international program that advocated a standard type and size of telescope, a wide-field 4-inch reflector called Moonwatch Apogee Scope. For the first half of the IGY these small telescopes provided the only consistent means of monitoring all the satellites. Later in the IGY more substantial Baker-Nunn cameras came into operation at several sites around the world. The Baker-Nunns were somewhat similar to the ballistic cameras at White Sands, but with tracking ability.

The Search for Small, Natural, Earth Satellites

The seed for an Earth satellite search was planted in Tombaugh's mind in 1943 when he and Henry Giclas attempted to recover asteroid 2101 Adonis (see Chapter 8). Since it had been lost, its location was very poorly known, and despite a search over a large area of sky, they did not find it. Its speed at closest approach was fast enough, however, that to search to its expected faint magnitude, the telescope was guided at the asteroid's expected rate by constant adjustment of the guiding eyepiece's micrometer reticle thread.[5] Despite the failure to find Adonis during that search, Tombaugh began to think about the possibility of searching more generally for satellites near the Earth. It was a logical extension of the trans-Saturnian search, only on the other side of the planetary fence.

Although Tombaugh's departure from Lowell and the early years at White Sands interrupted his observational work, the thought of another search never really left his mind. The idea of these possible objects goes back to William H. Pickering, who postulated them in 1923, but they were just some of many postulated objects that Pickering had suggested.[6] Tombaugh points out that Pickering had no idea how one might search for such objects.[7] In the early years at White Sands, as V2 rockets gave way to more sophisticated models and people began to dream about going into space, Tombaugh realized that not only was a search appropriate, but the army's Department of Ordnance Research might actually fund it.

Before we reached farther into space, especially with manned flights, it was important to know the obstacles awaiting people in that environment. Possible structural and medical problems had been studied, but what of the obstacles from without? Did the Earth have any natural satellites orbiting at distances close enough to the surface that they would never have been detected with telescopes? Tombaugh looked forward to a time when rockets much more powerful than what he saw at White Sands would be carrying people into space. He thought it prudent to search for obstacles in the way, particularly for a possible ring of small and faint rocks orbiting the Earth at distances close enough to avoid detection by normal astrophotography.

The goal was to search to about thirteenth magnitude, which could reveal a seventeen-foot-diameter object ten thousand miles away, or a 380-foot-diameter object at the Moon's distance. Satellites orbiting the planet would be searched for in 110 "distance-zones," ranging from ten thousand to one million miles from the center of the Earth.[8] The zones farthest out would be the hardest to search, since the cameras tracked at close to the sidereal rate (the rate at which stars move across the sky) and the wide fields were crowded with star images.

Serious discussions about the satellite search began in 1951, and by June 19, 1952, the proposal was complete. Including salaries and material, Tombaugh asked for only $19,000 a year. The proposal's main purpose was "one of basic research and interest to science." At the same time, it had military possibilities because procedures would naturally be developed to follow "extremely high altitude rockets."[9]

If an object was discovered, it would greatly assist geodesists in making better intercontinental surveys. For astronomers, a find could result in better astrometric surveys, as well as an improved understanding of the solar system. In the hopeful days of 1952 the proposal also suggested that comets "smaller than any now known to exist," as well as Earth-approaching asteroids, could turn up. The proposal's final purpose shows surprising foresight for a document of the time: the project could "aid as a base in the establishment of an artificial satellite or space station, if the characteristics of such a body are favorable."[10] Since a faint ring of rocks orbiting the Earth would definitely pose a danger to space vehicles, the Army Office of Ordnance Research agreed to fund the project.

It would take a camera moving across the sky at a rate much faster than sidereal to detect such objects. The cameras would track at rates as high as four hundred degrees per hour; that is, if the exposure were to last an hour, the telescope would move more than once across the sky and back to its starting point. The unique requirement was not so much in the camera or its mount but in the motor drive, which needed to support this enormous variation in drive rates. A small satellite at a distance of 26,180 miles from the center of the Earth would revolve in twenty-four hours; if the object were 5,000 miles distant, it would revolve in two hours.[11]

The cameras were of different types and sizes and had fields as large as thirteen degrees. The telescopes used were a Schmidt camera, a fast f/1.6 telescope with an 8½-inch correcting plate and a 12½-inch mirror—"a beautiful instrument"; and an Aero-Ektar K-24 lens. The rates were chosen so that a satellite at the required distance would be stationary on the plate. Tombaugh boasted that the equipment and strategy were good enough to detect "a clean white tennis ball" at a distance of one thousand miles.

The idea was to record objects within thirty degrees (each side) of the celestial equator and just outside the Earth's shadow, when they are sunlit and most easily detectable. Tombaugh arrived at the thirty-degree limit by reasoning that all but the most recently captured small objects would have been pulled to within ten degrees of the plane by the Earth's equatorial bulge, and that the objects farther out, where the pull of the Sun and Moon is stronger, would lie within a thirty-degree belt.[12] Because the telescope was generally moving so quickly, only the brightest stars were recorded on the plates, as long trails, thus easing the search.

Tombaugh found an elegant solution to the problem of how to confirm suspects. The third-plate concept he had used in his planet search was obviously inadequate for an object that would get away within minutes. The solution: during each exposure, the telescope's declination axis would be offset a tiny amount. The two parts of each five-minute exposure were kept deliberately uneven, so a five-minute plate would last possibly two minutes before the declination shift and three minutes after it. This way, if the short trail of an orbiting object was discovered, Tombaugh would know whether the telescope tracking was outrunning it or going too slowly. Assuming a circular orbit and using Kepler's laws of orbital motion, he could then recover the object. In December 1953 the program began at Lowell Observatory. Tombaugh devoted only half his time to this project during the first year; with the other half he was still directing the optical tracking. Claude Knuckles took most of the Schmidt photographs that first year. The satellite search involved brief flights to and from Flagstaff from the fall of 1953 until June 1956; each trip would last almost two weeks of each month, centered on the new moon of each lunation. During those years 13,450 photographs were taken,[13] containing hundreds of thousands of star trails. The process of examining each film for suspect satellite images took about fifteen minutes. The people involved in the project, including Tombaugh, each took turns at a scanning machine.

This new sojourn at Lowell put Tombaugh in a very different position from before. No longer the Kansas farm boy turned planet searcher, he returned to Lowell with a search project he had proposed and for which he had obtained support from the U.S. military. He brought his own assistants with him. Instead of the observatory struggling to get funds to employ Tombaugh, he was bringing funds to the observatory for a project he was directing.

In November 1956 the project had the opportunity to make a brief search for satellites of the Moon during one of the longest total lunar eclipses on record. A few nights before the eclipse, with Brad Smith and Chick Capen, Tombaugh began by setting up his 5-inch richest-field telescope on a field tripod, attaching a photometer to it, and then measuring the sky brightness at different times to calibrate the needed exposures on the 13-inch A. Lawrence Lowell telescope, which was to be the main instrument used. This way they had a precise indication of how long their exposures should be. On eclipse night the information from their own eyes was confirmed by the photometer: this eclipse was dark! They would be able to expose for fifteen or twenty minutes on their two main instruments, the 8½-inch Schmidt camera and the 13-inch astrograph. Guiding on one of the lunar craters, they made three exposures that precisely matched the Moon's motion during the eclipse. Because the eclipse allowed long exposures with the powerful 13-inch, this particular search would have located satellites as small as fifteen feet in diameter at the Moon's distance! None was found, but Tombaugh's hope, expressed in his 1956 *Interim Report* prepared before the eclipse, was obvious: "If [a satellite] is found, it will be the first great grandchild in celestial bodies known to astronomy." [14]

As the program continued to evolve, Tombaugh thought that observing from the equator would offer the best chance of detecting objects at the distance zones closest to the Earth. For this ambitious part of the program, site selection was a challenge. At first a site in Kenya was considered, but political unrest from the Mau Mau uprising scuttled that idea. Uganda was also a possibility, along with two spots in South America: Bogotá, Colombia, and Quito, Ecuador. After consulting a local meteorologist, Tombaugh found the weather prospects unsatisfactory at Bogotá. Quito had the triple advantage of almost straddling the equator, having satisfactory weather, and being at a high altitude of 9,300 feet. In November 1955 Tombaugh traveled to Quito to investigate the site and make arrangements for the search.

Moving the project to Ecuador presented an administrative headache for White Sands, an institution that had no mechanism for handling operations in other lands, and for months the arrangements went nowhere. While waiting for something to happen, Tombaugh thought that if the program was set up at New Mexico College of Agriculture and Mechanic Arts (now New Mexico State University), funding the Ecuador part would be easier, since the college handled international agricultural projects frequently. Moreover, the college's Physical Science Laboratory had been involved in many White Sands projects over the years, such as the reduction of ballistic data taken by Tombaugh's tracking cameras. When he contacted the Army Office of Ordnance Research to inquire if the program could not be managed just as well from New Mexico State, the answer was yes, and Tombaugh and his project moved over. Thus in the fall of 1955 Tombaugh resigned from White Sands.

In August 1956 Tombaugh made his second trip to Quito, this one to start the operational part of the search program. Scott Murrell and Jimmie C. Robinson joined him. In 1957 and 1958 Murrell and Robinson spent eight months in Ecuador, with good weather one year and poor the next. Observations were suspended during Quito's rainy seasons.

The Ecuador trips included two new members of the search team and a 12-inch reflector telescope that had been completed by the Fecker Company in August 1956. The telescope was used to carry two fast 4-inch-diameter Taylor-Hobson cameras, an 8-inch focal length (f/2) and a 6-inch focal length (f/1.5). Once the program was started, a new problem arose: with so many pictures necessary, how could one devise a system to change plates rapidly? Tombaugh's solution: attach film to the surface of a clean 4-by-5-inch glass plate. Plate holders were not used, and changing film was simple; after each exposure the film was removed and protected, another was slapped on, and a spring-backed lid was closed.

Because of its narrow field, the 12-inch was never used for actual photography. It was intended for pointing, guiding, and simply carrying the two cameras. Tombaugh had a twofold intention in purchasing this telescope. In Quito it served as a carrier for the two cameras, but if he later decided to start a planetary patrol program, such a 12-inch reflecting telescope would be ideal. "The army had a practice that when a project was done, there was a three-year transfer to a school that could use the equipment. The university got the equipment free of charge; you have to be a little scheming sometimes to get things done! I was getting to be a schemer."

In August 1956 Murrell and Robinson began to take photographs. There were so many zones, however, that photography of all of them became almost prohibitive. A change of strategy to visual searching was instituted. In this case, instead of simply using the 12-inch Fecker reflector as a mount, it was actually employed in the visual search with a 45-power Erfle eyepiece.[15] Its low-power field of fifty arc minutes (almost two Moon diameters) was complemented with the 3.5-degree field of ten-by-eighty binoculars. The instruments were pointed twelve arc minutes north of the zenith, and three-hour searches were held each evening beginning near the end of twilight, with additional searches before dawn.

As the 1957 season was about to close, the Soviets successfully launched Sputnik I. Tombaugh asked Robinson and Murrell to stay a while longer: "Don't come back yet; let's get some photographs of Sputnik!" Thus the project did reveal a satellite; it succeeded in getting some of the earliest photographs of the first artificial one. For that purpose the program was extended until March 1958.

In a symbolic sense that early photograph also marked the end of any practical search for natural Earth satellites. Today there are thousands of artificial satellites in all kinds of orbits, and any search carried out now would have the impossible task of distinguishing a natural satellite from the artificial additions. Objects as faint as thirteenth magnitude could be almost anything—small booster parts, fuel tanks, and all manner of "astronaut's glove" space junk. Tombaugh's project was beautifully timed, a natural outcome of the rocket work at White Sands, and can never be done again.

With no discoveries taking place in all that time, Tombaugh decided to close the project. "We felt we had learned enough; I didn't feel justified pouring more money into it because essentially I had learned my answer. . . . After taking about fifteen thousand films, plus the visual looking, we concluded that we could send rockets out in space with very little risk of collision with natural objects. Until we did that work, no one knew what to expect out there. Had these objects existed, they would never have shown up on plates" taken at the normal sidereal rates.

The project's final report appeared in June 1959. A few dozen suspects had turned up; none of them was confirmed. The report suggests that they were either film defects or, if real, small asteroids passing the Earth.[16] The project did yield much data on many bombardments of Earth—in the form of faint telescopic meteors.

Some of Tombaugh's assistants from that project later left their own marks in astronomy. Charles F. Capen started at White Sands, moved to Lowell in 1953 to assist with this search, and later became prominent as a Mars observer. Scott Murrell continued as the main observer for the planetary patrol at New Mexico State, and Bradford Smith went on to become imaging team leader for the Voyager I and Voyager II odysseys to Jupiter, Saturn, Uranus, and Neptune. Finally, the 12-inch Fecker telescope also went to research on the planets: Tombaugh's attempt to keep the telescope for his planetary patrol was successful.

Brad Smith had joined the search program somewhat surreptitiously. He had been assigned to the Army Map Service and was working for John O'Keefe, a geodesist who later proposed that the strange rocks called tektites had come from the Moon. They were doing geodetic survey work, the measuring of precise distances of land masses, for military purposes. At the time, they were using the Moon as an object for triangulation. By timing occultations of stars by the Moon from two different sites on two continents, O'Keefe's team would be able to determine distances between two land masses thousands of kilometers apart to an accuracy of a few meters. O'Keefe had heard of Tombaugh's work searching for natural satellites closer to the Earth and of some rumors that one had even been found. If so, O'Keefe's group would have used it to get much better triangulation than could be obtained with the Moon. Unfortunately, communication between Ordnance Research and the Army Map Service did not allow the latter to know whether Tombaugh had found an object or not. In late 1955 O'Keefe assigned Smith to join Tombaugh, assist in his work, and learn what he could.

After working with Tombaugh's project for a year, Smith retired from the army and had the option of moving to Massachusetts or staying with Tombaugh in Las Cruces. At the time, astronomy was strictly Smith's hobby; however, Mars was having a fabulous opposition that year. In a sense it was Tombaugh's enthusiasm for Mars that persuaded Smith to stay in Las Cruces and enter planetary science. The two scientists thus began a close friendship that would last for years.

Lowell Again

By 1952 Tombaugh was really beginning to miss planetary work. "I grew up on an eyepiece," he had said, and although he was working with optics at White Sands, he was out of astronomy. At the same time, Albert Wilson, Lowell's new director, asked Tombaugh to consider returning to the observatory he had left six years earlier. A survey for stars of high proper motion was being considered. The idea was that Tombaugh's long planet survey had recorded much of the sky and that by repeating the survey now, nearby stars themselves might move. Tombaugh was asked to visit Lowell, to take and blink a few plates.

On May 22, 1952, Tombaugh opened up the 13-inch to expose two plates. After processing, he placed one of the new plates into the comparator along with a plate he had exposed on March 16, 1936. This time he was searching not for planets but for nearby stars that would, over two decades, move very slightly across the fixed background of more distant stars. Once the old and new positions have been measured, the information can assist in calculating a star's distance.

The initial search revealed that the project was a "gold mine" of possibilities: on these plates, Tombaugh found five stars that had moved.[17] "I was interested to see how it would work out, and it was quite a revelation to see all these little things moving — different positions, different angles, different amounts. I felt like I lived in a dynamic universe." The highest number of proper motion stars consisted of red dwarf stars concentrated in the fourteenth and fifteenth magnitudes. If our part of the Galaxy is typical, this finding indicates that the Galaxy is full of small, very long-lived dwarf stars.

Meanwhile, at White Sands, life was beginning to lose its luster. One of the civilians "was trying to make it tough," and Tombaugh felt it prudent to return to Lowell at least to see if enough had changed. The new blinking work would be considerably eased with the purchase of a new blink comparator. Instead of an eyepiece, it used a projected field that was far more comfortable to study.

It had been a good summer. Tombaugh had taken plates, had done about three weeks of solid blinking, and had a chance to revisit his old home. The start of a formal proper motion survey was still uncertain. The role Tombaugh would play in the future, however, would be merely the blinking of all those plates from the proper motion survey. There seemed no prospect for serious planetary work, no prospect of a series of published papers. Tombaugh's answer was simple: " 'No, thanks,' I said, 'I have had all the blinking I want.' . . . I finally got it through my thick head that there was no place for me there anymore."

This decision, of course, was totally independent of the work he would do at Lowell for the Earth satellite search. That program had been proposed, with Lowell Observatory clearly designated as the site for the work, and as already noted Tombaugh would start observing there the next fall. Had Lowell offered him a position in planetary work, Tombaugh would have been tempted to take it. But the observatory was losing interest in continuing the planetary work E. C. Slipher had handled for so long and wanted Tombaugh only for his blinking experience. "I knew then I could not go back."

Some years later proper motion survey began under Henry Giclas's direction, and Robert Burnham, Norm Thomas, Charles Slaughter, and others worked on the project. The survey lasted longer than the original fourteen-year planet search. For each of the twelve thousand discoveries, the position angle and amount of drift were recorded, making this a much slower project in its analysis phase than the planet search. Since Flagstaff had grown in population and in street lighting in two decades, matching the new plates in magnitude and sky brightness with the old ones was a major problem. Eventually, the 13-inch was moved to the observatory's new Anderson Mesa site twelve miles southeast of Flagstaff where the sky was darker.

The requirements of the new survey were different from those of the old. Just one good plate of each region was required, instead of three. Although the plates needed to be taken at elevations in the sky that were similar to those of the old plates, so that differences in refraction would not be large enough to affect matching them up, the new plates did not have to be taken while the area was at opposition and highest in the sky in the middle of the night. Thus, plates could be taken of a region high in the sky at dusk or near dawn. Such times increased the chances for comet discoveries to occur on these plates, and Burnham, Slaughter, Thomas, and Giclas made six: Periodic Comet Slaughter-Burnham 1958 VI and Comets Burnham-Slaughter 1959 I, Burnham 1959 VII, Burnham 1960 II, Thomas 1969 I, and Giclas 1978 XXII.

Tombaugh, however, did not join this effort. With the end of the satellite search, and in the fall of 1958 the promise of another "great opposition" of Mars, Tombaugh ended this phase of his career. He turned once again toward the planets, not at Lowell, but at New Mexico State University.

Chapter 11

PROFESSOR TOMBAUGH

I T WAS time to return to the planets, but unlike his unfriendly departure from Lowell, this change in career would be pleasant and not abrupt. Tombaugh had already floated into New Mexico State to complete his Earth satellite program; he had an opportunity to transfer a 12-inch from that program to planetary use, and moving into planetary research seemed a natural way to come full circle. No matter how easy this change was, however, it had profound implications for Tombaugh. He now had much more time for planet drawings; in fact, he soon considered the visual component a necessary adjunct to the photographic patrol. That meant he needed to complete and mount his 16-inch reflector.

There was another big difference between 1946 and 1955. Tombaugh's record at White Sands was solid, and his first years at New Mexico State were productive; with his well-known accomplishments at Lowell, he was now highly respected. Tombaugh took this new step armed with two sets of ideas. The first was the tremendous amount of learning from E. C. Slipher, both in information about the planets and in techniques in drawing and photographing. The second was his experience at White Sands, which had persuaded him that one way or another, flights to the planets were approaching quickly.

There was also a sense of loss. Tombaugh knew that Lowell's planetary program was winding down—"what really hurt me was that E. C. Slipher had not taken a single photograph of Jupiter in two years. Here were all these changes and no record." There was a program at France's Pic du Midi, but in the absence of Lowell's, none in this hemisphere.

"It was time to get back into the planetary game," Tombaugh thought, and the sooner the better.

Since a nightly photographic patrol program involving multifiltered images of several planets had not been attempted to this complexity before, the early days of planning strategy were filled with discussions between Brad Smith and Tombaugh. By now the two had become very close. "Brad told us later that O'Keefe sent him down here as a spy to learn my methods, but I captured Brad!" Of course, this closeness did not prevent some sharp disagreements. Smith remembers how "we hardly agreed on anything. When each of us came up with a theory, the other was always looking for ways to shoot it down. That was very constructive. But it spilled into politics and everything else." When it came to science, "Clyde and I have never agreed on the respective values of visual observations and imaging. He always felt that high-quality visual observations were the best, because you could see more, and I agreed with him there. But I considered them not to be scientifically very valuable because you had to depend upon a particular individual's memory in jotting down what he saw."[1]

Philosophically, the approach that Tombaugh and Smith decided on was to observe each of five planets the way meteorologists observe the Earth—constantly and in as many ways as possible. The concentration would be on Jupiter, Saturn, Mars, Venus, and Mercury, each to be recorded through six filters from infrared (IR) to ultraviolet (UV). Depending on the filter used, the exposures would average two seconds for a planet like Mars at opposition, lengthening to thirty seconds for Saturn. A continuous monitoring program for planets was needed urgently for two reasons. The coming planetary flybys would sweep over a portion of a planet very briefly, but a monitoring program would cover the planets over a far greater time base. Second, any anomalous or puzzling events discovered during a flyby could be compared with this archival record.

Persuading agencies to fund this program was difficult. Despite the good early images, the first year of the fledgling program was funded to the rather small level of $25,000 from the National Science Foundation. The foundation, however, was unwilling to support the program further, so Tombaugh found air force funding for two more years. Finally, a proposal was approved by the National Aeronautics and Space Administration (NASA). "I had a lot of worries wondering why I did not get continued funding. I used to lie awake at night wondering about that. Oh, I had some rough years, I'll tell you." Tombaugh took responsibility for all the paperwork to get things started, from transferring the army's 12-inch reflector to funding the program's salaries and other expenses. In 1959 he announced that he would keep one of his best operators, Scott Murrell, "if I had to pay him out of my own pocket." Bringing that threat to a meeting in Washington, he came back with enough funds to keep the program alive a while longer. "I was feeling my way around," Tombaugh reminisces, "to see what was lying around the corner. If you look, then you get funded. And when the cuts came, we survived; others didn't."

During a critical part of this time, Brad Smith had to leave to assist O'Keefe's geodetic program. It was a difficult year for him; during one of the best Mars oppositions in history he was racing up and down the California coast timing occultations.[2] The experience did have an advantage; he met Robert Leighton and was introduced to his experiments on image stabilization. With his return to NMSU, Smith continued working in that direction.

The 12-inch Reflector

Finding a site for the 12-inch telescope was a curious challenge for a university situated in one of the finest rural locations in the United States. To save the costs of developing a fresh site right at the start, Brad suggested that his own backyard might do. With a few dollars for some concrete blocks, the telescope was thus erected on some private property on Las Cruces's northwest side.

Setting up such an unusual site arrangement was not very difficult, since the university never officially knew about it. Neither Tombaugh nor Smith ever signed insurance papers for the telescope; formalities like that simply did not get in their way. "When you see Clyde perched up on the scaffolding of his 16-inch telescope, hanging out and leaning over thin air, you get a feeling for how concerned Clyde was about it. Admittedly, he was more concerned about others than he was about himself."[3] The 12-inch took its first Mars pictures around 1958 from this backyard patrol station, its f/6.7 primary providing beautiful images. The telescope stayed there for about two years until a more permanent site on "A" Mountain (formally known as Tortugas Mountain) was ready.

One reason the program survived is the dedication of its participants; eventually, it employed thirteen people. Tombaugh and Smith both insisted that observations be taken every clear night without exception. "A lot of times you would jump in the vehicle and run up 'A' Mountain, and the seeing would be too poor. We worked weekends; we worked holidays; we'd schedule people to work Christmas and New Years." [4] The "A" Mountain site's pointed summit has some of the best seeing in the country and is protected from nuisance visitors by a Physical Science Laboratory facility at the base.

The quality of the 12-inch was so good that one Lowell astronomer admitted over dinner to Tombaugh: "You guys down there get better planetary detail with that twelve than we can get with the twenty-four!" Needless to say, Tombaugh was "tickled to death." With the dawn of the sixties and the early plans for planetary probes, planetary images were becoming increasingly important; the long struggle for funding was temporarily over. As the nation looked toward space, New Mexico State's planetary patrol was prospering.

Mariners

The first planetary probes headed for Venus. Since that planet offers so little through ground-based telescopes, Tombaugh saw the mission as a sort of necessary evil; Venus offered the least expensive way to dry-run a first planetary mission. Also, Mars was poorly placed for a probe. Early in the summer of 1962, Jet Propulsion Lab's Mariner I mission roared off its Cape Canaveral launch pad and plummeted into the Atlantic. The Earth-Venus window, the period during which a spacecraft could be most easily launched from one planet to the other, would not be open much longer, so preparations for Mariner II accelerated. In August 1962 the sister ship left the cape. During first-stage flight the booster started to roll unexpectedly. Had that continued, the mission would have failed. Straightening out as quickly as it had started rolling, the craft ended its powered phase and left Earth orbit for Venus. On December 14, 1962, the first *in situ* data from another planet streamed to Earth.

Although the fabulous success of Mariner II was greatly encouraging, Tombaugh was tensely awaiting the next two Mariner missions, which were funded for the 1965 opposition of Mars. Continuous planetary patrol was especially important for Mars because its strong winds are always changing its face, and the planet is never exactly the same from one observation to another. In 1963 Tombaugh attended a symposium on the exploration of Mars. The opening salvos featured four panel members, beginning with Willy Ley, who stated opinions about the value of manned exploration of Mars. During the question period Tombaugh stood up, introduced himself, summarized in a sentence or two his three and a half decades of Mars observing, and finally suggested that manned flight to the planets may be a far-off dream but is not necessarily the most efficient way in the near term. Unmanned probes would be far more cost-effective, safe, and valuable to science. After the first intermission the panel—now of five—reassembled; Tombaugh had been asked to join as a full member.

In 1965, the year of Mariner IV, Tombaugh did a long series of visual drawings using his backyard 16-inch. His years of observational data, as well as several years of patrol images, gave him some expectations of what the mission would find. It was disappointing that the surface looked so bleak: "MARS AT NOON: NO CANALS," one headline blared. But there were craters, a real surprise for many planetologists and a personal triumph for Tombaugh, who had predicted them fifteen years earlier.[5]

In another area Tombaugh's hunch was not correct. Occasional cloud patches that showed up suggested to him the presence of water vapor. The clouds Tombaugh saw affected only the blue-filtered images that bring out atmospheric details, and not the red images. The effects were fog or haze, he felt, that cleared up at times, so that sometimes the surface was visible with a blue filter and other times not. These seasonal variations were water-related, Tombaugh proposed. When later photometric data showed that the blue images were still getting to the ground, and the effect was likely one of changing contrast, Tombaugh was unconvinced. Much later the clear Mariner IX images proved it. "One of the things that was most difficult for me to give up was the idea that Mars had primitive vegetation. The color changes fitted their distinctive colors. . . . I would not be so bold as to say that Mars is totally devoid of life."

Operating by eye for accuracy and instinct for deduction, as some other scientists have accused him, is more art than science. Even if his ideas were not always correct, as an observer Tombaugh was a nonpareil. For years he disagreed with others about the correct place of Encke's division in Saturn's A ring. Others drew it in different spots, or not at all, while Tombaugh, using the 24-inch Lowell refractor, saw it consistently "in the middle of ring A."[6] This is the position first observed by James Keeler with Lick Observatory's 36-inch refractor in 1888,[7] and spacecraft images have finally proved it correct. Tombaugh was right about Encke's division, as he was right about Mars's craters.

The 24-inch Reflectors

Eventually, the quality of work done with the 12-inch must have impressed NASA, for funds were provided for a 24-inch reflector to be built by the Boller and Chivens Company. The 12-inch, with an f/66 secondary, Tombaugh admits, "was kind of a gamble." It had been originally planned as an f/60, but the distance to the planetary camera extended it slightly. The other secondary, f/20, was rarely used. The results it presented during its first years encouraged Tombaugh enough to gamble still further on an f/75 system for the 24-inch. He had considered an even greater gamble — f/90 — but felt the field curvature would be too great. He remembered Lampland's three secondaries for his 42-inch, giving focal lengths of 53, 80, and 150 feet. The longest one never worked well, although an f/120 has been tried successfully at the University of Arizona's Tumamoc Hill in Tucson.

Tombaugh and Smith debated the 24-inch parameters at first; Smith felt a prime focal ratio of four would be acceptable. Tombaugh had strong feelings for a longer f/5 telescope. The f/4, he felt, was too fast, and it may be a challenge to work the optics to diffraction limit. The people at Boller and Chivens, he recalls with a smile, "thought we were crazy to want these optics; they thought we were a bunch of nuts!"

The battle was partly over field. The faster system would provide a wider field for cometary studies. One solution was to use rapidly changing secondaries. Tombaugh discussed that with Boller and Chivens, and they came up with a device that allowed a flip-flop to change the secondary in just a few seconds. The system worked so well that absolutely no recollimating was necessary. Astronomers forced to wait while a crew uses a crane to replace the entire top part of the tube in order to change secondaries will appreciate that feature. The f/40 secondary was crucial in the telescope's successful recording of the breakup of Comet West in March 1976.

With the February 1967 opening of the 24-inch, the Planetary Patrol and Study Project entered its mature years, and as discoveries mounted, respect for the program rocketed. From ultraviolet filter photographs that revealed wind patterns, Brad Smith and Elmer Reese confirmed that Venus's upper atmosphere has a four-day rotation period, a finding first suspected by Charles Boyer in the Congo. Gordon Solberg's work led to the discovery of circulation patterns in Jupiter's Red Spot long before spacecraft images confirmed them. The archives now contain more than one million planetary images, of which almost half are of Jupiter.

Later Tombaugh designed a second 24-inch for stellar and extragalactic research. Setting up that program was a difficult process, for the site had to be far more distant from town to take advantage of the dark sky of southern New Mexico. The location eventually chosen was Blue Mesa, but identifying it as a good site involved repeated trips to its summit with small telescopes to check out the seeing conditions. There was no road; with his colleague James Cuffey, Tombaugh mounted the hill with camping equipment, telescope, and means for recording data, all on donkeys! The group also laid out the access route, independently discovering trails that had been used by Native Americans centuries earlier.

When Tombaugh retired in 1973, Brad Smith took over the program, although by this time he was heavily involved in the Mariner planetary missions. It was the only possible choice; having started with Tombaugh out at White Sands, Smith had joined him at NMSU in 1958 and had switched programs in graduate studies from physics to astronomy as soon as the department was launched. In 1972 he became its first Ph.D. The length of time wasn't important to either Tombaugh or Smith at the time; although Smith was a graduate student, he was also working closely with Tombaugh in the planetary patrol effort as well as becoming increasingly involved with the Mariner Mars missions. It was when Tombaugh's astronomy program was finally created that Brad became serious about getting his degree. "He was very good, very smart, very reliable. He was my right-hand man for about fifteen years. I could not have done all that without him; he was that valuable." The feeling on Brad's side was equally positive: "I certainly would never have gone on professionally in astronomy if it hadn't been for Tombaugh." He added, "On a scale of one to ten, working with Clyde would really be an eleven. Had I not run into Clyde, I have absolutely no idea what I would be doing right now. But the probability is that it would have nothing to do with space and astronomy."[8] Two years later, Smith left for the Lunar and Planetary Laboratory in Tucson, and Reta Beebe assumed responsibility for the planetary patrol program.

Meanwhile, Pluto was becoming popular again as a subject for research, and hence Tombaugh's role in its discovery was of interest. He was asked to write a chapter for a four-volume solar system encyclopedia by its editor, Gerard Kuiper.[9] Sometime after Tombaugh submitted his chapter, Kuiper happened to drop by for a visit. They talked about the chapter and its contents. "You know, Clyde," Kuiper said, "although your discovery of Pluto was a very important event, knowing how thoroughly you went through it, I am more impressed with what you did *not* find out there."

The 16-inch

The story of Tombaugh's 16-inch actually began in the late 1930s with the grinding of the large blank. Although the mirror was ground around 1944, a heavy teaching load, followed by his time-consuming early work at White Sands, prevented his completing the telescope for many years. In 1960 the telescope was finally mounted in the yard of his Las Cruces home. Its focus, more than thirteen feet long, required an unwieldy platform that enabled the observation of objects near the ecliptic, within two hours of the meridian.

For visual observing, Tombaugh still considers a 16-inch f/10 about the size limit and often has to stop it down to 12 inches. The result is a monstrous telescope that in some neighborhoods would have led to angry homeowner association meetings. (Las Cruces's response was to name an elementary school after him!) In 1966 the Tombaughs moved to a large new house at the southern perimeter of town, near the university horse farm (Clyde admits they make good neigh-bors).

In the late 1940s Tombaugh had acquired a very heavy equatorial mount for the 16-inch. He and Jed Durrenburger had to move this massive mount to the Tombaughs' new house. Tombaugh was using a two-wheeled cart and a long wooden plank down which, he supposed, the mount would gently slide. Jed was not at all confident of success, but Tombaugh assured him that "it will hold." As usual, Tombaugh's pockets were full of so much change, knicknacks, and other necessities that Jed called him "a traveling toolbox." They started moving the mount. A few seconds later it ran off on its own and started to plummet down the plank. The plank snapped and slammed into Tombaugh's thigh, leaving an impression there of everything in his pocket. The board then hit Jed across his chest. The mount landed safely, and quickly, on the ground and survived to lead a new career at the NMSU campus with the Las Cruces Astronomical Society telescope.[10] Besides the mount, the society's telescope uses a surplus ten-foot dome from one of the Askania cinetheodolites, and thanks to an idea by Brad Smith, the tube was once an Honest John booster.

The Professor

Tombaugh cannot literally lay claim to bringing astronomy to NMSU, since Professor Hiram Hadley taught a course in 1888, actually before Las Cruces College was even started. With many interruptions, astronomy courses were taught through the intervening years, especially by a Clarence Hagerty, who actually headed a department of astronomy around 1925.[11]

When Tombaugh arrived on campus in 1955 to complete his satellite search, he took an office in the library. The Planetary Group he later formed grew to five members in 1961, when he began teaching introductory astronomy, through the Department of Geography and Geology. Later he added Planetology.

Launching the department was a unique feat for Tombaugh, whose master's degree lacked the mathematics and physics to be competitive for a modern department. However, his never being department head probably had less to do with degree status than with his own anathema to administrative duties. "If Clyde had wanted to be head," former department head Herb Beebe claims, "the fact that he didn't have a

Ph.D. wouldn't have stood in the way." With the new department set up, Tombaugh's teaching load increased heavily and, with it, marking papers. "I was disappointed with students who didn't apply themselves and who wrote horrible exam papers," Tombaugh notes sadly, "but then there are those who are live wires, which makes you feel you've been rewarded and it's worth it after all."

Once the planetary program was well established, Tombaugh was approached by university vice-president William O'Donnell with an interesting request. Would Tombaugh begin work to set up a full doctorate-level graduate program? At first he was not at all sure; there were a hundred reasons why a doctoral program couldn't get launched, not the least of which was that Tombaugh himself had no Ph.D. He wanted a promise that "the rug wouldn't be taken" before the arrange-ment was complete. O'Donnell assured Tombaugh that the program would work — among other reasons, to "beat out the University of New Mexico at Albuquerque!" Tombaugh was honest about the expense. To do it right, to attract the best students, he would need several good staff members and at least a moderately large (meter class) telescope. A pitifully funded program would curl in on itself, and Tombaugh felt that trying to start one like that would not be worth his time and repu-tation. O'Donnell's mind was not closed by these concerns, and the vice-president urged Tombaugh to proceed. "I was under the gun to get this going and felt uneasy about it," Tombaugh notes.

One of the earliest — and politically touchiest — decisions was under which department to put the program. Traditionally, such programs go under physics, but Tombaugh was concerned with how much indepen-dence he would have within that large department. In addition, he did not want his section becoming too theoretical at the expense of prac-tical observation. Scientifically, Tombaugh wanted a program with a strong planetary emphasis, one that would work closely with the space probes. Feeling that he would have more independence in the two-man Geography and Geology Department, he turned in that direction. On a more personal note his childhood interest in astronomy had come from that direction.

For all these reasons, a hybrid Department of Earth Sciences and As-tronomy was set up, with a meteorology program included. In Septem-ber 1968 the Planetary Patrol and Study Project officially became part of the new Department of Earth Sciences and Astronomy, with

Brad Smith as director of planetary programs and Clyde Tombaugh as professor of astronomy and senior astronomer at the observatory. With at first only one geologist, Tombaugh helped out by adding a geology course to his astronomy teaching load. The arrangement was immensely successful. Geology quickly increased in popularity, and both sections expanded quickly. Nevertheless, the plan was that the astronomy section would eventually be independent, and that happened not many years later.

The Sixties at NMSU

By the end of 1968, planning for the new astronomy program was almost complete. The planetary patrol was beginning to make some headway. With NASA's planetary exploration gaining in intensity, and with proven good results from the 24-inch, the funding battles finally seemed to be over. Tombaugh looked forward to several years without funding worries, several years of concentration in science. He had assembled a good team. All this would have been a challenging but pleasant experience for Tombaugh had not the social turmoil of the sixties intruded into New Mexico State. One of the members of his team was a straight-A undergraduate physics student named Gordon Solberg, whom Tombaugh had met originally at the Las Cruces Astronomical Society. Solberg had a thorough understanding of the atmosphere of Jupiter and a strong appetite for research work. Tombaugh initially assigned him to reduce and analyze Jupiter images, using a Mann measuring engine. When they decided to computerize the operation around 1965, Gordon proved so proficient at computer programming that he eventually wrote all the software for measuring and reducing these Jupiter data.

Solberg's work reached its zenith with the publication of a paper reporting his discovery and analysis of a ninety-day oscillation in the longitude of Jupiter's Red Spot. "That was my hallmark paper," Solberg recalls. "It focused on that one thing. The other papers were mostly apparition reports on Jupiter from year to year."

After Solberg graduated, Tombaugh encouraged him to proceed to master's work, but he resisted, preferring to continue his involvement with the program as a university employee and assuming the title of junior astronomer. At the same time, like many other people in that generation, he was taking a hard look at something far removed from Jupiter, the Vietnam War and its increasing toll and questionable purpose.

Around 1968 the campus showed signs of increasing unrest. A campus hang-out called The Hut was attracting students after hours, as well as the worry of the administration, who put the place under surveillance with a camera attached to a nearby building. As the unrest increased, demonstrations began and got larger, and one was ended with tear gas.

In September 1968 a mysterious underground newsletter called *The Conscience* made its debut on campus. Its articles were antiwar, anti–university administration, and antigovernment. Although the paper became immediately popular with many students, its uncomplimentary remarks about the university president and its advocacy of radical change made it a headache for the administration. The first few issues were anonymously produced. The administration tried and failed to find *The Conscience*'s printing press and worked to locate the distributors; finally, university officials publicly called for the editors to reveal their identities. That winter an issue of *The Conscience* appeared with a masthead that responded to the administration's challenge: it identified Gordon Solberg as editor.

Led by its vice-president for research, Richard Duncan, the university decided to fire Solberg. Tombaugh was taken aback by the whole affair; he had had no idea that Solberg, his own employee, was involved. Concerned that the political unrest could torpedo the new astronomy program, Tombaugh pleaded with Solberg to stop his publication activity.

It was a tragic confrontation. By now Solberg's politics had eclipsed his astronomical work, and he declined Tombaugh's request. Tombaugh felt that his talented young employee was being insubordinate; Solberg thought he was being respectful but had to follow his conscience. Meanwhile, several senior officials were calling Solberg "Tombaugh's Troublemaker" or even "that communist who works for Tombaugh"; with tempers so frayed, Tombaugh had good reason to fear his cherished program was in trouble, although in retrospect other department members feel that the process of creating the department had gone too far for this episode to reverse it.

Adding to Tombaugh's fears was his distress that he would probably lose Solberg. Repercussions about how it was being handled had
badly damaged morale in the department. It was a terrible position for
Tombaugh; the vice president was angry at him for not acting strongly
against Solberg, while some other members were equally angry that he
did not act strongly for him.

In January 1969 Solberg met with Duncan, who implied that the
whole thing would be forgotten if Solberg resigned. Misinterpreting
Duncan's ultimatum, a few days later Solberg agreed to resign as editor
of *The Conscience*. Duncan quickly informed him that resigning meant
not from the paper but from his position under Tombaugh.

The firing was traumatic for the university as well as for Solberg
and Tombaugh. Solberg had been hired not by the university but by
Tombaugh; he was being paid not by the state but by Tombaugh's own
planetary patrol program funds. The dismissal raised a storm, as principal investigators all over campus protested that the university had no
right to fire someone in that circumstance. The American Civil Lib-
erties Union also entered the fray, contacting Solberg and offering its
assistance in a lawsuit against the university. Tombaugh was asked to
testify for both sides during the court battle and refused. "It's your
affair," he thought; "you've made a big issue out of a molehill." The
university finally settled for $6,750. Solberg, who now lives quietly
north of Las Cruces, keeping bees and writing excellent newspaper
articles, has never gone back into astronomy.

The Solberg affair added to the stress and fatigue Tombaugh was
beginning to feel at the time. His heavy teaching load was supplemented with repeated out-of-town trips for telescope site selection.

Two years later his physical health began to deteriorate. Around 1970
or 1971 Tombaugh decided it was time to move the 16-inch and its
massive mount. Dismantling it was a challenge, and he began to have
sharp pains. They grew intense, almost unbearable, as ribs started to
crack. For a time, lung cancer was suspected. But then a new stage
started with a slow and steady collapse of his spine. From standing up
very strong and straight, he gradually began to stoop over. His morale
was low. After several months, however, the pains grew less severe, and
aside from the stoop, he has recovered.

During this time Tombaugh continued his efforts to move the 16-
inch, now with a lot of help from his son, Alden. First its wooden

planks had to come off, and then platform and telescope were carefully moved, a process that took several days and required a flatbed truck. With returning health, Tombaugh faced a long and happy retirement. In a sense the most significant contribution he made to the department was his simple strength of being part of it. "Clyde has a positive attitude," Herb Beebe says. "He has been a breeze constantly blowing in the right direction." [12]

As a professor, Tombaugh attended almost every colloquium, adding flair with his strong presence and persistent questions. One time a scientist from California was enthusiastically describing his measurements of temperatures on Mars, relating them to seasonal change. Predictably, Tombaugh asked question after question. At one point the lecturer paused, turned to Tombaugh, and said, "Well, that's a good question. Tell me, are you an astronomer?" [13]

Chapter 12

LATER YEARS

W HEN a young student, Lori Kushner, was finishing a school project on Neptune, she asked the author for some help in explaining the circumstances of the 1846 discovery. On his next visit he brought a recording for her cassette recorder:

"Hello, Lori, this is Clyde Tombaugh, who discovered Pluto in 1930, and I hear you are interested in Neptune," the tape began. Lori looked up as if she had been struck by lightning, then listened to Tombaugh's message:

When I was young I read about the discovery of Neptune with great interest, and of course Herschel, who found Uranus, was one of my heroes. I never dreamed that I would find a planet. . . . In astronomy, mathematics is very important. You need all the math you can get, all the science, like chemistry, physics, and geology, because you depend on these sciences to understand astronomy. . . . You also must understand you will have to be very dedicated to this task. You have to have the perseverance to go through some hard lessons if you expect to achieve anything. You also should remember that a lot of work in astronomy is not glamorous like you think it is. A lot is just plain drudgery that gets little publicity.[1]

Lori turned off her recorder. If Tombaugh's unexpected lesson sticks, she will remember that the real fame in astronomy, as in anything else worth doing, comes from inside.

Early Retirement Years

At the end of the spring semester in 1973 Tombaugh retired from New Mexico State. He still attends almost daily lunches and colloquia there but has never tried to interfere with the department. As Herb Beebe recounts, "He left his hands off. He would react if I asked him questions, and not in a negative way; he did it sensibly."[2] Tombaugh remains active in the Las Cruces Astronomical Society, whose 12-inch telescope is housed in the south building at the Tombaugh Observatory on campus, opened in 1972. Naturally, he has continued working with telescopes. Of the thirty-six optical surfaces he has ground, from a 1½-inch Gregorian mirror to a 20-inch, the 9-inch and the 16-inch are his favorites. For deep sky observing Tombaugh uses his 10-inch f/5 reflector, whose mirror he ground on weekends during World War II. In 1985 he removed the motor from an old lawn-mower chassis and mounted the tube assembly onto it using a Dobsonian-style wooden alt-azimuth mounting. He wheels it around his backyard, dodging trees to find an opening to the cluster or galaxy he is looking for.

One of the happiest aspects of retirement is that Tombaugh has returned to astronomy for fun. His 16-inch reflector is used more and more often; any guest with half of Tombaugh's gumption would be invited to climb a ladder to a platform high above the ground. One night the author shared a look at Saturn. Expertly masking his sudden realization that he had acrophobia, the author concentrated on details within the rings and on that night of fine seeing saw some small dark markings that traversed the rings. "Have you ever seen the spokes in the B ring?"

"Oh, yes," Tombaugh answered, "I've seen them several times. The seeing has to be pretty good, seven [on a scale of one to ten] or better." Although the huge platform swayed slightly, the telescope it surrounded was mounted solidly. Unfortunately, Saturn was too high in the sky for normal viewing from the platform. Tombaugh's solution: place two wooden planks at the corners of the platform and a box on top of the planks. Climbing on top of the box would thrust the observer to the eyepiece, his eye gazing at Saturn and his mind forgetting that he was precariously perched eighteen feet above the ground.

The Smithsonian has approached Tombaugh twice with unusual requests. One was to give a lecture, not about Pluto, but about the work he had done in other areas, a rare request. The second was more personal: would Tombaugh kindly donate his 9-inch reflector to the Smithsonian for display in the "nation's attic"? This telescope had started Tombaugh toward the outer solar system many years earlier, and the request was an honor to the instrument's role in the nation's history of science. Answering as an observer, not a historian, Tombaugh would have none of that: "They can't have it yet—I'm still using it!" In fact, one of his many telescope projects in retirement was to add a triangular bar support between the mount and the top end of the long tube to give it more stability.

In recent years Tombaugh has kept busier than many people are during their working lives. "The chief thing is I've never had time to sit in a rocking chair," he boasts. He spends half of each working day in his office, which is piled so high with papers, books, journals, and magazines that there is no room to maneuver. He still lectures occasionally at New Mexico State and takes an interest in maintaining the two NMSU 24-inch telescopes, including cleaning their mirrors. In almost every major observatory, cleaning is strictly limited to professionally trained staff; at NMSU Tombaugh does it, even though university rules prohibit him from getting paid without an administrative routine that would hire him as a consultant.

Although Tombaugh has not begun any new research projects since retirement, he has developed an interest in wind power. His theory blows against the traditional windmill design with long blades; instead, Tombaugh designed a long horizontal windmill that acts like a combine's reel, with paddles perhaps sixteen feet long. Standing just a few feet above ground, the device would rest on two small mounts strategically placed on the tops of ridges where the wind was strong, instead of on the tall towers that characterize most windmill designs. Pivoted weights on the paddles would act as a self-regulating device to keep the fans reacting to the wind with a constant motion. The key to the design was its location: it had to be at ridge crests where the wind was usually strong.

Tombaugh loves strategy in action, whether it is planet hunting, designing a tracking telescope or windmill, or football:

> I can see the strategy on the field with the quarterback using his head for a given situation. Each type of play has various degrees

of risk—you lose a ball, you lose yards, you face an interception. If you've got too many yards to make, you pass on the third down; you don't try to rush because you won't make it and you will lose the ball. As you get down toward the other end of the field—within striking range of the goal—the other team tightens up pretty fast. And it's better to get the guy in the end zone and let him catch a pass there; then he gets a touchdown. If he catches a pass five yards from the goal line and then gets tackled, he doesn't make it; he may go on and never get to the goal line. If it's toward the end of the game and you're behind, then you take bigger risks, because you're going to lose the game anyway and you might as well take it.

On September 28, 1974, the NMSU Aggies played University of Texas at Arlington. As part of a tradition that involves faculty in football, the coach asked Tombaugh to join the team through lunch and the pregame briefing. Even though the game was going well, Tombaugh was surprised when the coach said, "Mr. Tombaugh, call a play." At first Tombaugh resisted. "Just call anything!" Tombaugh noticed a backfield man named Jim Germany who he thought was quite talented. "Send Germany through a right tackle," Tombaugh suggested. The call worked. Germany flew seventy-four yards to a touchdown, the Aggies won forty-two to fourteen, and Tombaugh was named honorary coach of the year.[3]

Out of the Darkness

A few years before Tombaugh retired from New Mexico State, his thoughts began to return to the small planet he had found four decades earlier. Interest had started to grow in Pluto again; *Sky and Telescope* had asked him to write some reminiscences for the thirtieth anniversary back in 1960, and since then the planet had become an increasingly popular research subject. When Tombaugh first began to write some notes about his discovery, he found that the pressure of teaching and other departmental duties was so great that he could not get enough quiet time to write, and the early years of retirement were still hectic enough that he didn't write very much.

In 1978 an editor contacted Tombaugh to suggest a book that would tell the story of Pluto from two directions. Tombaugh would write the chapters dealing with Lowell Observatory and his own experiences there, and the noted British astronomy writer Patrick Moore would produce background sections about the asteroids, Uranus and Neptune, and other related subjects.

In March 1979 Tombaugh began a process of writing that took nine months of full-time work, complete in longhand. Finally, it was finished. For months Tombaugh heard nothing, and then a revised version came back. Unfortunately, a new editor had been assigned to the project; she told Tombaugh that the material would be heavily edited.

In *Out of the Darkness* Tombaugh wanted to do more than tell a simple story; through his writing he was presenting an observer's philosophy. It was the way Tombaugh expressed this point of view that the editor objected to, saying it would all have to be changed. Tombaugh was livid. The Pluto story had become his heart and soul, and he insisted that he be allowed to tell it his own way or not at all. "She completely rephrased everything. Took the liberty to write whatever she wanted." It was part of Tombaugh's personality to let an episode like this get the better of him. "I nearly went out of my mind," he complained. "I got so frustrated I resolved I'd never write another book."

1980: The Fiftieth

Work on *Out of the Darkness* was still keeping Tombaugh busy at the start of the celebration of the fiftieth anniversary of the discovery of Pluto, possibly the busiest year of his life since the discovery. It began with a celebration at the Adler Planetarium on January 21, half a century after the first of the three Delta Geminorum plates were taken. Discovery day itself was commemorated on February 18 with a major celebration at New Mexico State, at which Tombaugh received the university's highest honor, the Regents' Medal. Announcement day, March 13, was spent at Flagstaff.

Between these special meetings Tombaugh was rushing to complete revisions to *Out of the Darkness* in hopes that it could appear during that fiftieth anniversary year. By now all the difficulties had been resolved, and Tombaugh was enjoying the work. Late in the spring the illustrations had been selected, the final revisions had been made, and the book was ready for production.

On July 4, complete with a fifty-gun salute, Tombaugh was inducted into the White Sands Missile Range Hall of Fame for his work in optical tracking and ballistic data. Among others inducted that day were Werner von Braun and Frank Hemingway. A week later Tombaugh was in Arizona again, this time for a Western Amateur Astronomers banquet. For this meeting, he produced his "Ten Special Commandments for a Would-Be Planet Hunter":

Behold the heavens and the great vastness thereof, for a planet could be anywhere therein.
Thou shalt dedicate thy whole being to the search project with infinite patience and perseverance.
Thou shalt set no other work before thee, for the search shall keep thee busy enough.
Thou shalt take the plates at opposition time lest thou be deceived by asteroids near their stationary positions.
Thou shalt duplicate the plates of a pair at the same hour angle lest refraction distortions overtake thee.
Thou shalt give adequate overlap of adjacent plate regions lest the planet play hide and seek with thee.
Thou must not become ill at the dark of the moon lest thou fall behind the opposition point.
Thou shalt have no dates except at full moon when long-exposure plates cannot be taken at the telescope.
Many false planets shall appear before thee, hundreds of them, and thou shalt check every one with a third plate.
Thou shalt not engage in any dissipation, that thy years may be many, for thou shalt need them to finish the job![4]

In September *Out of the Darkness* finally appeared, an immediate success. In the first five years Tombaugh autographed 1,381 copies, "equal to a stack of books eighty-seven feet high." Finally, late in October the University of Kansas dedicated the Tombaugh Observatory, which houses his master's degree telescope. After feeling somewhat inferior for many years after the discovery, Tombaugh realized that year that he was regarded "with a considerable amount of respect. Prior to that

I thought they didn't like it. I misjudged the attitude of astronomers. I thought I was a nobody. I thought they had contempt for me."

Religious Thought

From his early Illinois days, when he read through the entire Bible, religious study—not religious dogma—has played an important part in Tombaugh's life. His interest in the Old Testament stemmed from its being the story of a group of people who survived under conditions of great stress. Like a number of scientists, Clyde and Patsy are active in the Unitarian Church in Las Cruces, and also like many scientists, they reject the illogical. Angered by those who would persecute Galileo and Giordano Bruno, who would plunder the laboratory of chemist Joseph Priestly, Tombaugh sees the current conservative tide in this country as a return to those days. His anger at dogmatic thought increased during his trips to Ecuador, where he saw the absolute poverty of the rural citizens. "The Church has such a hold on them! They've got a terrible population explosion there—social conditions are terrible—and the Church will not permit birth control? A God who makes childbirth painful because of the sin of one woman? . . . And even here, they want creationism taught in schools besides evolution theory. Why don't they include all the theories, then, like those of the Navajo tribe? They have valid traditions, too."

Tombaugh's religious thinking dates to his youth during the debates his family had about the Scopes evolution trial. Religious dogma, as opposed to religious thought, is "totally in disagreement with the findings of science." Tombaugh recalls a classmate's visit when he lived in Kansas; they were both sitting on a rock patio his father had made. The rocks were filled with simple marine fossils. "Can't you see that this country was under the sea at one time?" Tombaugh asked his friend. "Look at these fossils!"

"Oh, God just put them there to test our faith."

Tombaugh was astonished. "Is God to be a deceiver, like Satan?"

Tombaugh's view that religion cannot survive the logic of science does not excuse those who live a religion of science. Axiom: The mass of evidence supports the Big Bang. "I frankly do not believe the Big Bang theory," Tombaugh says of the single most accepted theory of the origin of the Universe. Concerned that scientists are coming to believe in it almost by faith, he asks, "Why should the Universe start from a point, essentially a point?" If religion has not changed much in Tombaugh's life, science surely has. He began by looking at planets through small refractors and doing simple sketches. Now he has attended several planetary encounters by spacecraft, including Voyager's fabulous visits to the four outer planets. "I lived in a day and age when you could know everything there was to be known about a planet like Mars or Jupiter. You can't now. . . . I feel overwhelmed, but I'm glad to see it."

"Life has always been worth living," Tombaugh reflects. "If I had to live my life over, I don't know if I would elect to do it much differently. I was very fortunate in every way; I had the opportunity and took advantage of it. All those hours, tedious as they were—I have never regretted them." [5]

Tombaugh's New Career

Thanks to the growing interest in Pluto, when Tombaugh retired, his "fan mail" was increasing dramatically. During the forties and fifties the Pluto story was virtually ignored; now Las Cruces loves Tombaugh's presence. A real-estate brochure advertises the fact that the discoverer of Pluto lives in Las Cruces, and in 1990 the Clyde Tombaugh Elementary School was opened on the city's south side. By the mid 1980s, however, intense publicity had become the most time-consuming part of his many activities; Tombaugh estimates that he has received about thirty thousand letters since 1930, as well as an average of two to three hundred visits a year. Although he generally answered those who had thought to enclose return envelopes with stamps, "it's been a horrendous problem. A lot of them I don't answer because I'm just getting tired of buying stamps for people who want information out of me for nothing." Tombaugh complained also that most of these questions could easily be answered by an encyclopedia, if only the person would care to look. Years of answering this mail has embittered Tombaugh somewhat: "It's getting a bit old, I tell you; it's getting old."

Tombaugh's second publicity problem was with the press, which could be counted on to contact him whenever someone reported or even postulated a promising new planet. "If I meet someone I size up as being capable of writing a good story, then fine and dandy, I'll help them; otherwise they're not writing about me! Because they'll get it all wrong, and I'm tired of making corrections to stories."

In 1986 Bernard McNamara, of the NMSU Astronomy Department, thought that the department could take advantage of Tombaugh's enormous popularity by raising the funds for a new postdoctoral fellowship through a private fundraising campaign. A high-profile effort would give national publicity to the department. After some discussion, the Clyde Tombaugh Scholars Fund was launched. A central focus of the fundraising was the amateur community, whose various clubs across the country would plan lecture tours in which large crowds would hear the story of how Pluto was discovered. In addition to an admission charge, an autographed certificate of participation and a poster would be sold.

Tombaugh's health posed a haunting question at the beginning. He had celebrated his eightieth birthday that year. Would he be able to stand what could develop into a grueling series of cross-country journeys and a long schedule of lectures? These fears were nullified after the project's first few months. In a three-day trip to the Riverside Telescope Makers Conference in May 1987, Tombaugh's spirits surged as he recalled his earlier years to an audience of more than a thousand amateur astronomers. As publicity for the campaign spread, Tombaugh was booked in cities all over the country.

On a northeastern tour in the fall of 1987, Tombaugh's schedule was intense. In one day he delivered a lecture in Philadelphia and a banquet talk in Bethlehem, appeared on a local radio program, autographed almost one hundred posters, and answered countless questions. The following day he appeared at an executive lunch, and then he toured an amateur observatory site. That evening he was the guest at a fundraising party for amateur astronomers. In a setting reminiscent of children meeting Santa Claus, the amateur astronomers spent several hours talking with the only living discoverer of a major planet. They saw Tombaugh as one of them, a role he plays well and honestly. A reception the following day was held at the headquarters of the American Association of Variable Star Observers and was also sponsored by *Sky and Telescope* magazine. Once again Tombaugh signed posters and recounted his discovery.

By the summer of 1988, the project was so well established that the almost monthly trips were becoming routine. Tombaugh was still balking at some, however. The trip to Stellafane Observatory, for example, would occur near the height of one of the century's best Mars apparitions, and at first he resisted. Finally relenting, he and Patsy made the three-day trip to Vermont that August, and in two days of work he raised more than seven thousand dollars. The crowd of almost three thousand people welcomed him in a beautiful outdoor setting. Overlooking the hillside was the old clubhouse that had inspired the *Scientific American* article Tombaugh had read in 1925. Sixty-three years later, this audience gave him a standing ovation before he said a word. As the applause ended and Tombaugh began his words under a twilight sky, the years melted away, the large crowd fading into a small group of people setting up telescopes to observe from the vastly different world of 1925. He was an amateur astronomer then. With passing decades Clyde William Tombaugh would use his experienced eyes to discover, his agile hands to construct, his fertile mind to teach, and his life story to inspire. In his amateur roots that August night, the discoverer had come home.

Appendix

A COMET AND A NOVA

A BIOGRAPHY sometimes leads its writer in unexpected directions. The present book revealed from glass plates a new comet and a nova.

While reviewing the objects Tombaugh had found during his planet search, I noted that he had discovered one comet. Not being aware of any Comet Tombaugh, I asked him about it. "I never reported it," he answered, "because it had been found on plates taken more than a year earlier. The comet would have been long gone." Further, it turned out that when he asked the Sliphers about it at the time, they agreed that reporting it at that late date would be pointless.

Having an interest in comets, I decided to try to uncover this filmy visitor. Although Tombaugh could not remember when in his search he found the comet, it had to have been sometime between 1929 and 1945. "But you shouldn't have too much trouble finding it," he added, "because I made careful notes for each plate on its envelope." All I had to do was examine all his notes and find the plate envelope with the comet image.

Ready for a little adventure, I left for Flagstaff in February 1986 and met Brian Skiff in Lowell Observatory's subterranean plate vault. I shuddered when I entered that room. One long wall was completely covered with several shelves of large white envelopes, each containing a plate. The amount of data on that wall was astounding. Elsewhere were plates taken with other instruments. The date of my visit was February 9, 1986, the perihelion of Halley's Comet, and I found plates of that same comet, taken on its previous return in 1910.

185

The old envelopes (with Tombaugh's meticulous notes) no longer protected these plates. Since acid in these covers had threatened to damage the plates, the observatory staff had replaced them with modern archival envelopes. However, the inscribed backs of the originals were preserved in a filing cabinet.

I reached for a stack of these notes and set to work. They were very well organized, with numbers of suspects, asteroids, variable stars, galaxies, and other objects carefully recorded. Although each envelope I checked said "no comets," the other things of interest were so engrossing that I spent several hours that afternoon going through the treasure. At one point Lowell's director, Art Hoag, walked by: "A convenient way to observe: comet hunting by reading!"

Finally, I reached the envelope for plate 327. Although the exposure had been taken on January 10, 1931, Tombaugh had not examined it until March 11 and 12, 1932, and he noted the probability of one comet. Brian and I carefully removed all three plates of this series and carried them to the comparator room. The instrument in place now is not the original one Tombaugh used, but the more recent model intended for the proper motion survey. It took some searching before we located the object, but with its diffuse appearance and small tail, there seemed no question that Tombaugh had really discovered a comet.

Not only was the twelfth magnitude object on all three search plates, but it was also at the edge of one of the Cogshall plates taken of another region—that camera had so much greater area coverage than the 13-inch that there was a very large overlap of regions. Leaving the comparator room that night, I looked down the hall of the old building where so many years ago Tombaugh had walked toward Slipher's office to report a new planet. The object embedded in tonight's plates was a lot closer to Earth than Pluto ever would be, and a lot less important, but nevertheless I felt as though I had rescued a filmy image that had been frozen in emulsion for fifty-six years.

Next we placed each plate into a digital scanning machine, a device that measures the positions of objects relative to those of "fixed" stars. Because the comet's image from the Cogshall camera was weak, we then had four positions: three accurate and one not so accurate. Before going further, Brian made a routine check of the *Astronomische Nachrichten*, where positions of newly discovered objects were published at that time. Incredibly, he found that the object had in fact been reported—by Carl Lampland and Kenneth Newman—but as an asteroid, not a comet! It had even been given a tentative designation, 1931 AN, in the tradition of asteroids, not comets. Probably, Newman had incorrectly identified the object's nature, measured its position very approximately, and reported it along with some asteroids. It is unlikely that Lampland checked that work, for he would surely have made the correct identification, and since Tombaugh was away at the University of Kansas at the time, he was not involved in the report. Armed with this information, Brian and I sent the following note to Brian Marsden of the International Astronomical Union's Central Bureau for Astronomical Telegrams, with the help of Ted Bowell:

Research by David H. Levy, University of Arizona, for a biography of Clyde W. Tombaugh has revealed the existence of a comet on plates taken with the 0.33m A. Lawrence Lowell Astrograph at Lowell Observatory for the trans-Neptunian planet search. Because the image had been found on fourteen-month-old plates, Tombaugh had decided not to report it at the time, and fifty-five years later he could not recall when the comet was discovered. This comet was, however, the only one Tombaugh noted in his entire planet search. Subsequent searches in the Lowell plate archives by Levy and Brian A. Skiff revealed images of this comet on three plates taken January 1931 and examined on March 11 and 12, 1932. Three more images were located on plates taken simultaneously [with the Cogshall camera] as well as a fourth taken a day earlier near the edge of an adjacent field. [At this point several accurate positions are provided.] The comet was diffuse, with strong condensation, with tail at least 2 arcminutes pointing west. Subsequent inquiry by Skiff revealed that approximate positions for this object were published in *Astronomische Nachrichten* 249:105 as 1931 AN.

The following day we talked with Brian Marsden. In the case of comets observed long ago, it was the IAU's policy not to give designations unless definitive orbits could be obtained. The three good positions were enough for only a very rough preliminary orbit. Unless more images could be found, there would be no announcement of Comet Tombaugh.

Other work kept me from getting back to the problem until the summer of 1986, when I visited Harvard College Observatory's plate archive. One of the largest collections of astronomical photographs in the world, this archive represents years of regular patrol work. Armed with predicted positions for the comet based on the preliminary orbit, amateur astronomer Peter Jedicke and I went through the several plates of the region that the archive had from 1931. The search was frustrating. Plates taken with Harvard's narrow-angle cameras did not happen to record the correct region, and the wide-angle plates that did cover the star field did not show the comet because it was too faint.

In November 1986 I visited the Heidelberg Observatory during a Halley's Comet symposium. Brian Marsden had suggested that the comet might appear at the very edge of a plate they had. With Steve Edberg, of the International Halley Watch, I retrieved the plate and compared its star images with a map. The comet, alas, was just off the edge of the plate. We did, however, look at some of the beautiful old telescopes at Heidelberg, including the refractor with which Max Wolf had recovered Halley's Comet in 1909.

A few days later I visited Meudon Observatory near Paris. Unfortunately, their plate collection also did not show the comet. As time passed, I managed to check the plate collections of other observatories, either personally or by asking for help from the astronomers in charge. W. Wenzel, of the Sonneborg Observatory, identified some plates of regions that included the comet, but like the Harvard plates, they did not record objects as faint as the comet. Other observatories checked— Bamberg, Lick, and David Dunlap—did not have images, although Mt. Wilson did have a plate taken within a degree of where the comet had passed.

After a year of searching I abandoned my hope of finding more images. The evidence I now have is sparse: a comet did pass through Lynx in the winter of 1931, reaching at least twelfth magnitude for a time. It is probably a new comet, never before identified, although it could be a return of an unknown periodic comet.

Was Tombaugh right in his decision not to report the comet at the time? Today communication among observatories is so efficient that the search for year-old images would not be as difficult as it was then. At the time, the decision seemed appropriate. In any event, too few observatories were taking wide-area coverage, and the comet was never seen again.

This search project gave me a firsthand view of the meticulous care with which Tombaugh examined his plates. The many objects he found, mostly asteroids, are still being studied from the plates by Ted Bowell, Brian Skiff, and others, and the harvest of information will probably continue in the years ahead. Since Tombaugh had said that he found only one comet, I stopped looking for additional comets on his plate notes after those of 1934. Some time later Brian Skiff uncovered a second Tombaugh comet, this one reported on a September 1931 plate envelope. Brian measured its positions and reported its presence to the Central Bureau.

Now wondering what else might remain buried on Tombaugh's trans-Saturnian search plates, I visited Lowell again in the summer of 1987 and checked every one of the plate envelope notes. Now the story was happier—on March 22, 1931 (March 23 Universal Date), Tombaugh had discovered a nova, a star undergoing an outburst, in the constellation of Corvus. That object was also never reported officially. His plate notes: "I nova suspect. 'T 12' near southwest corner of plate, magnitude about 12, confirmed well on 5" Cogshall plate of March 22. No trace of object on 13-inch plates of March 20 and 17, 1931. The image is exactly deformed, like the other star images in the neighborhood. Evidently a very remarkable star to rise from 17 or fainter to 12 in 2 days time. Position: Epoch 1855 RA (1855.0) 12h 13m Dec -17 deg 40' or RA (1930.0) 12h 16'.9 Dec -18 degrees 05'. This object was discovered on May 25, 1932, at 11:00 AM."[1]

Since this distant star does not go wandering about the sky as comets do, I decided that it might be worth further investigation. It did not appear on the earlier plates, which recorded objects as faint as seventeenth magnitude in blue light, but it did appear as a twelfth magnitude star on the night of March 23 UT both on the 13-inch telescope plate and on the Cogshall plate exposed at the same time. A star of about eighteenth magnitude does appear at the nova's exact position on a plate taken on March 7, 1954, with the large Schmidt for the Palomar Observatory Sky Survey.

On September 11, 1989, I once again visited the plate stacks at Harvard College Observatory. In three days of work I examined the nova's position on more than 260 patrol plates, spanning a period from 1930 to 1988, and found the star in outburst nine additional times.

It appears that Nova Corvi 1931 is a cataclysmic variable, a "dwarf nova" whose periodic outbursts can easily be observed today. In November, when Corvus started to appear in the morning sky after its conjunction with the Sun, I began daily checks of the star's region on every clear morning with my 16-inch reflector as well as with occasional Schmidt camera and CCD images. For several months, as the region became more conveniently visible, I saw nothing unusual.

On March 23, 1990, fifty-nine years to the day after the initial outburst, I looked through the 16-inch and saw a magnitude 13.6 star where nothing had appeared before! Within an hour Skiff had informed Rob McNaught, who arranged for a spectrum using the Anglo-Australian Telescope, and Gary Rosenbaum arranged for spectra to be taken immediately from Kitt Peak. Robert Kennicutt and Dennis Zaritsky obtained a spectrum with Steward Observatory's 90-inch telescope, and Richard Henry obtained two on successive nights using Kitt Peak's 84-inch. The following morning Circular 4983 was issued from the International Astronomical Union.

Although the personal story pauses here, the scientific story has just begun. Steve Howell, who is studying the spectra, says that they indicate that the star was unusually hot during its outburst and that it was changing characteristics quickly. It appears as though Tombaugh's latest object shares the mystery and curiosity aroused by his other discoveries.[2]

NOTES

Chapter 1. Looking into Chapman's Homer

1. C. Tombaugh, plate notes for No. 171 (courtesy Lowell Observatory).
2. C. Tombaugh, interview, November 8, 1985.
3. C. Lampland, personal diary, February 18, 1930 (courtesy H. Giclas).
4. C. Tombaugh, plate notes for No. 171 (courtesy Lowell Observatory).

Chapter 2. The Educated Ones

1. W. Shakespeare, *Twelfth Night* II, v, 131–34.
2. R. Tombaugh, interview, May 1, 1989.
3. Ibid.
4. Ibid.
5. C. Tombaugh, "Stars above a Kansas Farm," *The Furrow* (Jan.–Feb. 1963): 2–3.
6. C. Tombaugh and P. Moore, *Out of the Darkness: The Planet Pluto* (Harrisburg, Pa.: Stackpole, 1980), 18.
7. R. Tombaugh, interview, May 1, 1989.
8. Tombaugh and Moore, *Out of the Darkness*, 18.
9. Tombaugh and Moore, *Out of the Darkness*, 19.
10. I. Rath, *Boy Planet Seeker* (Dodge City: Rollie Jack, 1963), 75.
11. E. Tombaugh Spreen, *On to Kansas* (private memoir, 1989), p. 8.
12. Tombaugh and Moore, *Out of the Darkness*, 20.
13. Ibid., 19.
14. C. Tombaugh, letter to Napoleon Carreau, April 26, 1926 (courtesy E. O. Mannery).

15. Tombaugh and Moore, *Out of the Darkness*, 20.

16. Ibid., 21.

17. Ibid.

18. G. Kronk, *Comets: A Descriptive Guide* (Hillside, N.J.: Enslow, 1984), 289. Although this comet returns every six years, perturbations in its orbit have resulted in its being far less impressive. In May 1989, for instance, the author observed it almost fourteen magnitudes fainter.

19. For a modern discussion of the mirror-making process, see Richard Berry, *Build Your Own Telescope* (New York: Scribner's, 1985), chap. 10.

20. E. Tombaugh Spreen, 10.

21. W. Hoyt, *Planets X and Pluto* (Tucson: University of Arizona Press, 1980), 179.

22. B. Marsden, personal communication, September 1989.

23. Y. Vaisala, "Minor Planet Work at the Astronomical Observatory of the Turku University," *Turku Informo No. 6* (1950): 12.

Chapter 3. Uranus, Neptune, and Planet X

1. C. Lubbock, *The Herschel Chronicle: The Life-Story of William Herschel and His Sister Caroline Herschel* (London: Cambridge University Press, 1933), 15.

2. Ibid., 60.

3. Ibid., 66.

4. W. Herschel, "Account of a Comet," in *The Scientific Papers of Sir William Herschel*, vol. 1, 30–31, ed. J. L. E. Dreyer (London: Royal Society and Royal Astronomical Society, 1912).

5. Lubbock, *Herschel Chronicle*, 95.

6. J. Ashbrook, *The Astronomical Scrapbook* (Cambridge, Mass.: Sky Publishing, 1984), 44.

7. M. Grosser, *The Discovery of Neptune* (New York: Dover, 1979), 117.

8. Ashbrook, *Astronomical Scrapbook*, 329.

9. W. Hoyt, *Planets X and Pluto* (Tucson: University of Arizona Press, 1980), 120.

10. Ibid, 133.

Chapter 4. The First Year

1. W. Hoyt, *Lowell and Mars* (Tucson: University of Arizona Press, 1976), 268ff.

2. J. Bossidy, Toast, Holy Cross Alumni Dinner, 1910.

3. W. Hoyt, *Planets X and Pluto* (Tucson: University of Arizona Press, 1980), 146.

4. In 1989 the new trustee, William Putnam, arranged for this walk to be paved with concrete. The positions of the Sun and the nine planets were inscribed on it, with appropriate informational signs planted alongside. Pluto was right next to the 13-inch dome. When Tombaugh walked this path one more time during the dedication ceremony, he remarked nostalgically how nice it would have been if the concrete had been there since 1929!

5. C. Tombaugh and P. Moore, *Out of the Darkness: The Planet Pluto* (Harrisburg, Pa.: Stackpole, 1980), 99.

6. Ibid.

7. Ibid., 100.

8. Hoyt, *Planets X and Pluto*, 160.

9. Tombaugh and Moore, *Out of the Darkness*, 106.

10. A. Norton, *Norton's Star Atlas*, 3d ed. (London: Gall and Inglis, 1921); 18th ed. (Essex: Longman's, 1989).

11. Hoyt, *Planets X and Pluto*, 133.

12. Tombaugh and Moore, *Out of the Darkness*, 111.

13. Hoyt, *Planets X and Pluto*, 184.

14. Ibid., 182.

15. C. Tombaugh, interview, January 31, 1986.

16. Hoyt, *Planets X and Pluto*, 183.

17. Tombaugh and Moore, *Out of the Darkness*, 117.

18. Ibid., 119.

19. C. Tombaugh, "My Early Years at Lowell Observatory," unpublished manuscript, 2.

20. C. Tombaugh, "The Discovery of Pluto," in *Source Book in Astronomy, 1900–1950*, ed. H. Shapley (Cambridge: Harvard University Press, 1960), 72.

21. Tombaugh and Moore, *Out of the Darkness*, 118.

22. Tombaugh, "Discovery of Pluto," 71.

23. R. Burnham, Jr., personal communication, June 1967.

24. Tombaugh and Moore, *Out of the Darkness*, 121.

25. Ibid., 101.

Chapter 5. Discovery's Wake

1. D. Berger, personal communication, May 1986.

2. B. Marsden, personal communication, September 1987.

3. C. Lampland, personal diary (courtesy H. Giclas).

4. Ibid.

5. Ibid.

6. W. Hoyt, *Planets X and Pluto* (Tucson: University of Arizona Press, 1980), 193–94.

7. A. J. Whyte, *The Planet Pluto* (Toronto: Pergamon Press, 1980), 34.

8. C. Tombaugh and P. Moore, *Out of the Darkness: The Planet Pluto* (Harrisburg, Pa.: Stackpole, 1980), 131.

9. Lampland, diary.

10. Ibid.

11. C. Tombaugh, interview, August 17, 1987.

12. Lampland, diary.

13. *Harvard Announcement Card* 108, March 13, 1930.

14. V. M. Slipher, *Lowell Observatory Observation Circular*, March 13, 1930.

15. Ibid.

16. Lampland, diary.

17. C. Tombaugh, personal communication.

18. Lampland, diary.

19. *Allentown Morning Call*, March 14, 1930, p. 1., col. d.

20. Hoyt, *Planets X and Pluto*, 87, 120, 131.

21. R. Tombaugh, interview, May 1, 1989.

22. Ibid.

23. Lampland, diary.

24. Hoyt, *Planets X and Pluto*, 204–5.

25. W. Wisniewski, personal communication, November 1986.

26. B. Marsden, personal communication, September 1989.

27. Hoyt, *Planets X and Pluto*, 210.

28. F. Whipple, personal communication, April 1989.

29. Hoyt, *Planets X and Pluto*, 210.

30. Ibid., 213.

31. Ibid., 217.

32. *Encyclopedia of Animated Cartoon Series*, ed. J. Lenburg, New York: Quality Paperbacks, 1983. Courtesy S. Fisher and P. Manly.

33. Hoyt, *Planets X and Pluto*, 216.

34. Lampland, diary.

35. Hoyt, *Planets X and Pluto*, 249.

36. R. Tombaugh, interview, May 1, 1989.

Chapter 6. The Thirties

1. W. Hoyt, *Planets X and Pluto* (Tucson: University of Arizona Press, 1980), 250.

2. C. Tombaugh, plate envelope notes for No. 1122, plate taken September 22, 1936 (courtesy Lowell Observatory).

3. G. Kronk, *Comets: A Descriptive Guide* (Hillside, N.J.: Enslow, 1984), 127–28.

4. C. Tombaugh, plate envelope notes for No. 410, June 1932 (courtesy Lowell Observatory).

5. C. Tombaugh, plate envelope notes for No. 419, July 1932 (courtesy Lowell Observatory).

6. C. Tombaugh, plate envelope notes for No. 487, August 25, 1933 (courtesy Lowell Observatory).

7. D. Eicher, *Deep Sky Observing with Small Telescopes* (Hillside, N.J.: Enslow, 1989), 184.

8. J. L. E. Dreyer, *A New General Catalogue of Nebulae and Clusters of Stars* (London: Royal Astronomical Society, 1888).

9. C. Tombaugh, plate envelope notes for No. 417, June 1932 (courtesy Lowell Observatory).

10. C. O. Lampland and C. W. Tombaugh, "Object NGC 5694, A Distant Globular Star Cluster," *Astronomische Nachrichten* (August 1932). Nr. 5888, Band 246.

11. *NGC 2000.0*, ed. R. Sinnott (Cambridge, Mass.: Sky Publishing, 1988).

12. A. Ingalls, ed., *Amateur Telescope Making, Advanced* (Scientific American, 1937), 638–41.

13. D. Alter, *Lunar Atlas* (1964; rpt., New York: Dover, 1968).

14. F. Edmondson, 13-inch photographic record book (courtesy Lowell Observatory).

15. J. Edson, personal communication, July 12, 1989.

16. W. Haas, personal communication, August 14, 1989.

17. P. Lowell, *Mars* (1895; rpt., Paul Luther, 1978), 131.

18. W. Hoyt, *Lowell and Mars* (Tucson: University of Arizona Press, 1976), 204.

19. C. Tombaugh, interview, September 28, 1985.

Chapter 7. Galaxies

1. A. Berry, *A Short History of Astronomy from Earliest Times through the Nineteenth Century* (1898; rpt., New York: Dover, 1961), 338.

2. Ibid., 339.

3. T. Webb, *Celestial Objects for Common Telescopes* (1917; rpt., New York: Dover, 1962), 22.

4. H. Shapley, *Through Rugged Ways to the Stars* (New York: Scribner's, 1969), 79.

5. C. Tombaugh, "The Great Perseus-Andromeda Stratum of Extra-Galactic Nebulae and Certain Clusters of Nebulae Therein as Observed at the Lowell Observatory," *PASP* 49, 291 (1937): 259.

6. H. Shapley, *The Inner Metagalaxy* (New Haven, Conn.: Yale University Press, 1957), 35.

7. Tombaugh, "Extra-Galactic Nebulae," 263.

8. E. Hubble, *The Realm of the Nebulae* (1936; rpt., New York: Dover, 1958), 69.

9. Ibid., 72.

10. C. Tombaugh, plate envelope notes for No. 487, August 24, 1933 (courtesy Lowell Observatory).

11. H. Shapley, *Galaxies*, 3d ed., rev. P. Hodge (Cambridge, Mass.: Harvard University Press, 1970), 191–92.

12. C. Tombaugh, interview, July 9, 1988.

13. G. Abell, *Exploration of the Universe*, 4th ed. (Philadelphia: Saunders, 1982), 628–40.

Chapter 8. War and Departure

1. C. Tombaugh and P. Moore, *Out of the Darkness: The Planet Pluto* (Harrisburg, Pa.: Stackpole, 1980), 168.

2. Ibid., 171.

3. C. Tombaugh, "Reminiscences on the Discovery of Pluto," *Sky and Telescope* 19, 5 (March 1960): 270.

4. N. Bowditch, *American Practical Navigator* (Washington, D.C.: U.S. Government Printing Office, 1943).

5. C. Cunningham, *Introduction to Asteroids* (Richmond, Va.: Willmann-Bell, 1988), 96.

6. *Minor Planet Circular* 4541, *IAU Circulars* 3012 and 3041.

7. C. Tombaugh, interview, August 17, 1987.

Chapter 9. White Sands Years

1. A. Vick, interview, July 8, 1989.

2. Ibid.

3. J. Edson, personal communication, July 12, 1989.

4. Ley idea from G. H. Stine, July 10, 1989.

5. W. Haas, personal communication, August 23, 1989.

6. Edson, op. cit.

7. J. Marlin, interview, July 8, 1989.

8. F. Hemingway, interview, July 8, 1989.

9. J. Gold, interview, July 8, 1989.

10. F. Hemingway, interview.

11. P. and C. Tombaugh, interview, May 14, 1985.

12. J. Gold, interview.

13. F. Hemingway, interview.

14. J. Durrenberger, interview, July 9, 1989.

15. G. H. Stine, interview, July 10, 1989.

Chapter 10. *Astronomy at White Sands*

1. W. Baade, *Evolution of Stars and Galaxies* (1963; rpt., Cambridge, Mass.: MIT press, 1975), 100–101.

2. C. Tombaugh, "A Survey of Long-Term Observational Behavior of Various Martian Features That Affect Some Recently Proposed Interpretations," *Icarus*, 8, 2 (March 1968): 235.

3. C. Tombaugh, "Geological Interpretation of Martian Features," *Journal of the Association of Lunar and Planetary Observers*, 4, 10 (October 1950): 4–6.

4. J. Gold, interview, July 8, 1989.

5. C. Tombaugh, *Interim Report on Search for Small Earth Satellites* (Las Cruces: New Mexico College of Agriculture and Mechanic Arts, 1956), 3.

6. W. H. Pickering, *Popular Astronomy* 31, 23 (1923).

7. Tombaugh, *Interim Report*.

8. C. Tombaugh, *Search for Small Satellites of the Earth* (project proposal), 2.

9. Ibid.

10. Ibid.

11. Tombaugh, *Interim Report*, 3.

12. Tombaugh, *Search for Small Satellites*, 2.

13. Tombaugh, *Interim Report*, 19.

14. Ibid., 9.

15. C. Tombaugh, J. C. Robinson, B. A. Smith, and A. S. Murrell, *The Search for Small Natural Earth Satellites: Final Technical Report* (Las Cruces: New Mexico State University Physical Science Laboratory, 1959), 66.

16. Ibid., 21.

17. C. Tombaugh, plate envelope notes for No. PM-3, May 22, 1952. Courtesy of the Lowell Observatory and Brian Skiff.

Chapter 11. Professor Tombaugh

1. B. Smith, interview, November 21, 1988.

2. Ibid.

3. Ibid.

4. S. Murrell, interview, August 14, 1989.

5. C. Tombaugh, "A Survey of Long-Term Observational Behavior of Various Martian Features That Affect Some Recently Proposed Interpretations," *Icarus* 8, 2 (March 1968): 235.

6. A. Alexander, *The Planet Saturn* (New York: Dover, 1962), 409.

7. Ibid., 254.

8. B. Smith, interview, autumn 1987.

9. C. Tombaugh, "The Trans-Neptunian Planet Search" in *The Solar System*, vol. 3, G. Kuiper and B. Middlehurst, eds., (Chicago: University of Chicago Press, 1961), chap. 2.

10. J. Durrenberger, interview, July 9, 1989.

11. H. Beebe, Department of Astronomy history (New Mexico State University, 1988).

12. H. Beebe, interview, August 14, 1989.

13. W. Haas, interview, August 14, 1989.

Chapter 12. Later Years

1. C. Tombaugh to L. Kushner, December 1987.

2. H. Beebe, interview, August 14, 1989.

3. Game statistics courtesy of B. McCann, Sports Information Department, New Mexico State University, Las Cruces.

4. C. Tombaugh, "Ten Special Commandments for a Would-Be Planet Hunter," July 1980.

5. C. Tombaugh, interview, March 23, 1987.

Appendix: A Comet and a Nova

1. C. Tombaugh, plate envelope notes for March 22, 1931 (courtesy Lowell Observatory).

2. D. Levy, S. Howell, T. Kreidl, B. Skiff, and C. Tombaugh, "The Historical Discovery and Recent Confirmation of a New Cataclysmic Variable in Corvus." *Publications of the Astronomical Society of the Pacific*, in press.

BIBLIOGRAPHY

Abell, G. *Exploration of the Universe*, 4th ed. Philadelphia: Saunders, 1982.

Alexander, A. *The Planet Saturn*. New York: Dover, 1962.

Alter, D. *Lunar Atlas*. 1964. Reprint. New York: Dover, 1968.

Ashbrook, J. *The Astronomical Scrapbook*. Cambridge, Mass.: Sky Publishing, 1984.

Baade, W. *Evolution of Stars and Galaxies*. 1963. Reprint. Cambridge, Mass.: MIT Press, 1975.

Berry, A. *A Short History of Astronomy from Earliest Times through the Nineteenth Century*. 1898. Reprint. New York: Dover, 1961.

Bowditch, N. *American Practical Navigator*. Washington, D.C.: U.S. Government Printing Office, 1943.

Cunningham, C. *Introduction to Asteroids*. Richmond, Va.: Willmann-Bell, 1988.

Dreyer, J. L. E. *A New General Catalogue of Nebulae and Clusters of Stars*. London: Royal Astronomical Society, 1888.

Eicher, D. *Deep Sky Observing with Small Telescopes*. Hillside, N.J.: Enslow, 1989.

Grosser, M. *The Discovery of Neptune*. New York: Dover, 1979.

Herschel, W. "Account of a Comet." *The Scientific Papers of Sir William Herschel*, ed. J. L. E. Dreyer. London: Royal Society and Royal Astronomical Society, 1912.

Hoyt, W. *Lowell and Mars*. Tucson: University of Arizona Press, 1976.

——. *Planets X and Pluto*. Tucson: University of Arizona Press, 1980.

Hubble, E. *The Realm of the Nebulae*. 1936. Reprint. New York: Dover, 1958.

Ingalls, A., ed. *Amateur Telescope Making, Advanced*. Scientific American, 1937.

Kronk, G. *Comets: A Descriptive Guide*. Hillside, N.J.: Enslow, 1984.

Kuiper, G., and Middlehurst, B., eds. *The Solar System*. Chicago: University of Chicago Press, 1961.

Lampland, C. Personal Diary. February and March 1930. Courtesy H. Giclas.

Lampland, C. O., and Tombaugh, C. W. "Object NGC 5694, A Distant Globular Star Cluster." *Astronomische Nachrichten*, Nr. 5888, Band 246. August 1932.

Lenburg, J. *The Encyclopedia of Animated Cartoon Series*. New York: Quality Paperbacks, 1983.

Lubbock, C. *The Herschel Chronicle: The Life-Story of William Herschel and His Sister Caroline Herschel*. London: Cambridge University Press, 1933.

Mayall, N. U. "Edwin Hubble, Observational Cosmologist." *Sky and Telescope* 13, 3 (January 1954): 85.

Norton, A. *Norton's Star Atlas*. 3d ed. London: Gall and Inglis, 1921. 18th ed. Essex: Longman's, 1989.

Rath, I. *Boy Planet Seeker*. Dodge City: Rollie Jack, 1963.

Shapley, H. *Galaxies*. 3d ed. P. Hodge, rev. Cambridge, Mass.: Harvard University Press, 1970.

——. *The Inner Metagalaxy*. New Haven, Conn.: Yale University Press, 1957.

——. *Through Rugged Ways to the Stars*. New York: Scribner's, 1969.

Smith, H. "A Brief Report of Photographic Observation of Mars during the Present Opposition." *Publications of the Astronomical Society of the Pacific* 51 (December 1939).

Spreen, E. Tombaugh. *On to Kansas*. Private memoir, 1989.

Tombaugh, C. "The Discovery of Pluto." *Source Book in Astronomy, 1900–1950*, H. Shapley, ed. Cambridge: Harvard University Press, 1960.

——. "Geological Interpretation of Martian Features." *Journal of the Association of Lunar and Planetary Observers* 4, 10 (October 1950): 4–6.

——. "The Great Perseus-Andromeda Stratum of Extra-Galactic Nebulae and Certain Clusters of Nebulae Therein as Observed at the Lowell Observatory." *Publications of the Astronomical Society of the Pacific* 49, 291 (1937).

——. *Interim Report on Search for Small Earth Satellites*. Las Cruces: New Mexico College of Agriculture and Mechanic Arts, 1956.

——. Plate Envelope Notes. Courtesy Lowell Observatory.

——. "Reminiscences on the Discovery of Pluto." *Sky and Telescope* 19, 5 (March 1960), 264–70.

——. "Stars above a Kansas Farm." *The Furrow* (Jan.–Feb. 1963).

——. "A Survey of Long-Term Observational Behavior of Various Martian Features That Affect Some Recently Proposed Interpretations." *Icarus* 8, 2 (March 1968), 227–58.

——. "The Trans-Neptunian Planet Search." In G. Kuiper and B. Middlehurst, *The Solar System*, vol. 3. Chicago: University of Chicago Press, 1961.

Tombaugh, C., and Moore, P. *Out of the Darkness: The Planet Pluto*. Harrisburg, Pa.: Stackpole, 1980.

Tombaugh, C.; Robinson, J. C.; Smith, B. A.; and Murrell, A. S. *The Search for Small Natural Earth Satellites: Final Technical Report*. Las Cruces: New Mexico State University Physical Science Laboratory, 1959.

Vaisala, Y. "Minor Planet Work at the Astronomical Observatory of the Turku University." *Turku Informo No. 6* (1950): 12.

Webb, T. *Celestial Objects for Common Telescopes*. New York: Dover, 1962.

Whyte, A. *The Planet Pluto*. Toronto: Pergamon Press, 1980.

INDEX